Moving Theory
into Practice

Moving Theory into Practice

Digital Imaging for Libraries and Archives

ANNE R. KENNEY | OYA Y. RIEGER
EDITORS AND PRINCIPAL AUTHORS

RESEARCH LIBRARIES GROUP
MOUNTAIN VIEW, CALIFORNIA

All URLs cited in this book were verified December 1999–
January 2000.

Research Libraries Group
Mountain View, California 94041-1100

Printed in the United States of America

This work is cataloged in the RLG union catalog (RLIN® database) with the record ID CRLG00-B4.

Research Libraries Group Cataloging Data

Kenney, Anne R.
Moving theory into practice : digital imaging for libraries and
archives / Anne R. Kenney, Oya Y. Rieger, editors and principal
authors. — Mountain View CA : Research Libraries Group,
2000.
x, 189 p. : ill. ; 28 cm.
Includes bibliographical references and index.
ISBN 0-9700225-0-6
1. Library materials—Digitization.   2. Archival materials—
Digitization.   3. Image processing—Digital techniques.
4. Digital preservation.   I. Rieger, Oya Y.   II. Research
Libraries Group   III. Title.   IV. Title: Digital imaging for
libraries and archives.

BOOK DESIGN AND COMPOSITION: BookMatters, Berkeley,
California
COVER DESIGN: tompertdesign, Palo Alto, California
PRINTER: Sheridan Books, Ann Arbor, Michigan

CONTENTS

ILLUSTRATIONS

Figures

Plates

ACKNOWLEDGMENTS

MANY INDIVIDUALS have supported the writing of this book. We are particularly grateful to Robin Dale of RLG for all her hard work and invaluable insights in making this book a reality. She was heavily involved in all phases of this effort—from conceptualization to reviewing and editing to managing production. Her encouragement when we needed it most enabled us to deliver the manuscript within a very tight timeframe.

We have had an outstanding cast of contributing authors. We particularly would like to acknowledge the great work of the guest chapter authors: Paula de Stefano (Preservation Department, New York University Libraries), Carl Lagoze and Sandra Payette (Department of Computer Science, Cornell University), John Price-Wilkin (Digital Library Production Service, University of Michigan), and Peter Hirtle (Cornell Institute for Digital Collections).

We extend our appreciation to those who reviewed the individual chapters in draft form and provided useful comments and suggestions: Howard Besser (University of California Los Angeles), Carl Fleischhauer (Library of Congress), Janet Gertz (Columbia University Library), David Levy (Xerox PARC), Patti McClung, Steven Puglia (National Archives and Records Administration), Thornton Staples (University of Virginia), and John Weise (University of Michigan).

The diversity of perspectives and backgrounds presented in the sidebars greatly strengthen this book. Many thanks to Arnold Arcolio (RLG), Ross Atkinson (Cornell University Library), Meg Bellinger (Preservation Resources), Elizabeth Z. Bennett (JSTOR), Bruce H. Bruemmer (University of Minnesota Libraries), Peter D. Burns (Eastman Kodak), Ross Coleman (University of Sidney Library), Paul Conway (Yale University Library), Kenn Dahl (Prime Recognition), John P. Eakins (University of Northumbria at Newcastle), Carl Fleischhauer (Library of Congress), Franziska S. Frey (Rochester Institute of Technology), Georgia Harper (University of Texas System), Margaret Hedstrom (University of Michigan), Diane I. Hillmann (Cornell University Library), Geri Bunker Ingram (University of Washington Libraries), Frank Klaproth (State- and University Library Göttingen), Drew Lathin (STS Design Consultants, LLC), Norbert Lossau, (State- and University Library Göttingen), Clifford Lynch (Coalition for Networked Information), Deanna B. Marcum (Council on Library and Information Resources), Richard Marisa (Cornell Information Technologies), Carla Rickerson (University of Washington Libraries), Beth Sandore (University of Illinois), Deborah Skaggs (Frank Russell Company), Abby Smith (Council on Library and Information Resources), Sarah Thomas (Cornell University Library), Janet K. Vavra (Inter-University Consortium for Political and Social Research), Bruce Washburn (RLG), Don Williams (Eastman Kodak), Don Willis (Connectex, LLC), Deborah Woodyard (National Library of Australia), Yecheng Wu (Able Software Corp.), and Susan M. Yoder (RLG).

We would like to thank our friends and colleagues in the Cornell University Library, Preservation and Conservation Department: John Dean, director, and Barbara Berger Eden, reformatting librarian, for reviewing each and every chapter, and also for their personal support of our work; Rich Entlich, digital projects researcher, provided valuable background research and editorial assistance; Mary Arsenault, Carla DeMello, and Allen Quirk also helped at various stages.

We end with our continuing thanks to the Research Libraries Group for joining with Cornell University Library in this enterprise and for publishing this book. We especially wish to acknowledge Liz Chapman's assistance as copyeditor. Her careful and relentless editing of the book is very much appreciated—even when it meant thousands of our carefully chosen words ended up on the cutting-room floor. This book is much better for her effort.

Introduction:

MOVING THEORY INTO PRACTICE

Anne R. Kenney and Oya Y. Rieger

DIGITAL TECHNOLOGIES and the ever increasing popularity of network access have irrevocably changed the information landscape. Today, the World Wide Web supports more sites than the Library of Congress holds books, and the number of network users rivals the number of library users and newspaper readers combined. Governments, higher-education institutions, and the entertainment and commercial sectors are quickly developing the technological infrastructure and the content to make digital access ubiquitous, reliable, and fast as light. The Internet will become the agora for research, teaching, expression, publication, and communication.

These developments have had profound impact on cultural institutions of all types. For several decades, libraries and archives have relied on computing and connectivity to streamline technical operations, improve existing services and provide new ones, and, increasingly, deliver full content. At the same time, they continue to husband physical collections, including material that is also available over the Internet. Straining resources of all kinds, libraries and archives end up lagging on both fronts. They are losing the paper chase and in the digital realm many, especially the young, see them as the information provider of last resort. The Benton Foundation's 1996 report "Buildings, Books, and Bytes" found that Americans aged 18–24 were the least enthusiastic about libraries in the digital age.[1] Today, many expect all relevant information

to be online and to have customized, granular access to it.[2]

Clearly libraries and archives have to change if they are to continue to serve as society's primary information providers. Much of the effort is devoted to the major challenge of managing "born-digital" information. But an impressive number of institutions have also turned to creating digital surrogates from retrospective resources. A May 1999 survey by the Association of Research Libraries (ARL) revealed that 91% of responding institutions are conducting or planning digitization of special collections holdings.[3]

Why this interest in digitization? Peter Lyman and Brewster Kahle refer to the process as creating "born-again" digital materials—and in that term lies more than just a play on words.[4] We believe that cultural institutions are investing in digitization for two reasons: First, they remain convinced of the continuing value of such resources for learning, teaching, research, scholarship, documentation, and public accountability. Second, they recognize that changing user behavior may jeopardize these resources and their stewardship.

Several years ago, the National Archives and

1. "Buildings, Books, and Bytes: Libraries and Communities in the Digital Age," *Benton Foundation,* www.benton.org/Library/ Kellogg/buildings.html.

2. "Top Technology Trends: New Tech Trends Update from the 1999 ALA Annual Convention," *LITA (Library Information and Technology Association),* www.lita.org/committe/toptech/ trendsmw99.htm.

3. Preliminary survey results reported at the spring 1999 membership meeting of ARL directors, which focused on "Special Collections in the Digital Age."

4. Peter Lyman and Brewster Kahle first coined this term in "Archiving Digital Cultural Artifacts: Organizing an Agenda for Action," *D-Lib Magazine* (July/August 1998) www.dlib.org/dlib/ july98/07lyman.html.

Records Administration (NARA) adopted a new strategic plan with the mission "to ensure ready access to essential evidence . . . that documents the rights of American citizens, the actions of federal officials, and the national experience."[5] Ready access has for many come to equal network access, with use and ease of use correlating closely. To

breathe new life into older cultural resources requires institutions to respond to such change. Not coincidentally, the government's settlement with the tobacco industry included the creation of a public Web site with the crucial documentation of health-related effects of smoking.[6] The government recognized the imperative to make information available in whatever form users require. And indeed it seems such effort pays off in promoting access: institutions that make retrospective resources available online report that use of the digital surrogates dwarfs use of their physical counterparts.

WHY THIS BOOK?

Moving Theory into Practice is intended as a self-help reference for libraries and archives that choose to retrospectively convert cultural resources to digital form. It advocates an integrated approach to digital imaging programs, from selection to access to preservation, with a heavy emphasis on the intersection of cultural objectives and practical digital applications. Certainly, a plethora of information on digital imaging is readily available, much of it on the Web. Indeed, we see less need to emphasize the general and the technical than we did when Cornell published *Digital Imaging for Libraries and Archives* in 1996. At the end of this chapter, we offer some references to basic information and ways to keep up with changes in the technology. The principal aims of *Moving Theory into Practice* are to foster critical thinking in a technological realm and to provide librarians, archivists, curators, administrators, technologists, and others working in cultural repositories with the means to move beyond accepted theoretical constructs to implementation strategies that reflect distinct institutional missions and capabilities.

Consider these real-life scenarios:

▸ Institution A receives a large multimillion-dollar grant from a private foundation to undertake a major digital imaging initiative. Project staff negotiate with a company to provide a full turn-

Real-Life Choices
Abby Smith
Director of Programs
Council on Library and Information Resources

In an ideal world, decisions about digitization would begin with careful thought about your target audience and what materials would work well in digital form for a variety of potential uses. This process would be accompanied by wide consultation with reference librarians, collection development specialists, preservation experts, information technology specialists, and, of course, representative users. Preservation staff would carefully assess the physical condition of the items to be scanned and do the necessary repair or stabilization. Catalogers would create finding aids or bibliographical records to render those digital items quickly retrievable in a robust database that links easily and effortlessly to other databases of similar digital collections.

That may well happen sometime in the future and somewhere over the rainbow, but today, back here in Kansas, decisions about what to scan, for whom and why are usually shaped by broader factors. Most digital projects are driven at least in part by strategic institutional considerations that often conflict with the well-ordered and narrowly confined parameters of an ideal digital conversion project. This usually happens when digitization is seen as serving institutional priorities above and beyond providing resources to a specific targeted audience, such as undergraduates or art historians. Often, a library or university administration will develop digitization projects to position the institution for development of infrastructure and capacity, for fundraising, or simply because it is "the thing to do." None of these reasons are bad per se and they are intrinsic to an institution large and ambitious enough to undertake digital conversion of its holdings. There are really only a few bad reasons to undertake digital projects. In general, these bad reasons stem from those involved not understanding or not honestly telling key decision makers just what digits can and cannot do.

Bad reasons for digitizing include:

▸ *Preservation:* Digitization is not cheaper, safer, or more reliable than microfilming. Unlike a frame of high-quality microfilm, a digital image is not a preservation master, even if it is referred to as a digital master. At present, the only direct way that digital reformatting contributes positively to preservation is when the digital surrogate reduces physical wear and tear on the original or when the files are written

CONTINUED ON PAGE 3

5. National Archives and Records Administration home page, www.nara.gov.

6. The Tobacco Institute home page, www.tobaccoinstitute.com/default.htm.

key solution. Six months later, the company declares bankruptcy and because the system is proprietary, the project cannot be migrated to a new platform. Institution A quietly abandons the effort after one year.

▸ Institution B purchases a planetary, flatbed scanner to scan 18th-century bound volumes. Product literature claims that the system can produce grayscale images. Using the equipment, staff discover that the grayscale capability is actually a halftoning process, poorly suited to the content of the volumes.

▸ Institution C outsources its scanning to a company that provides imaging services to a range of demanding customers in government and industry. The company can produce an acceptable product, but has never established a full quality assurance program. Institution C discovers this only after signing the contract. The project is delayed until acceptable inspection procedures can be developed and implemented.

▸ Institution D conducts a series of digital imaging projects throughout the 1990s, financed by outside funds. In 1999, it discovers that the jukebox acquired five years earlier is malfunctioning and the manufacturer no longer supports the product. The optical disks are not readable on any other machine. Nearly a third of a terabyte of information is at risk. A new storage solution has to be found quickly—as well as the money to pay for it, because the institution has not earmarked funds for ongoing maintenance.

These experiences illustrate vividly how research institutions need to make informed decisions about the viability and use of digital imaging technology. The impulse to embrace things digital is strong, but too often infrastructure—costs, personnel, systems, and preservation—gets insufficient thought and delivery falls short of the promise.[7] Information professionals can little afford to make mistakes in initiating and maintaining digital

programs. They must assess carefully the pros and cons of technology choices in a cultural context. The best way to ensure good decisions is to become a knowledgeable consumer of the technology.

CONTINUED FROM PAGE 2

to computer output microfilm that meets national standards for quality and longevity.

▸ *Collection management*: A digital surrogate of analog material can never replace the original item and should never be forced to, no matter how much shelf space you think you can gain by chucking the analog originals. If an institution decides to deaccession its brittle newspapers to gain space, it had better secure microform copies of the papers it wishes to serve its patrons. Nor does digitization result in cost savings; at least, it has not to date. Scanning items and creating records demand big up-front expenditures and the long-term maintenance of digital assets comes at a price. Digitization enhances existing services and hence adds costs. As something new and desirable, the increased cost may still be well worth the investment.

Good reasons for digitizing collections include:

▸ *Access*: Digitization can allow wider access to fragile and rare materials. It can allow researchers to use materials from anywhere on the face of the planet, allowing physically disparate collections to exist in the same virtual space. It can even turn massive numbers of analog monographs, novels, magazines, and serials into searchable databases, allowing a completely new type of access to those texts.

▸ *Professional development*: Digital projects allow staff to build a critical set of skills the best way possible: by doing. They can offer unprecedented opportunities for developing qualified and motivated staff.

▸ *Influence building*: Even today, digital projects are usually experimental and permit a rare and precious let's-try-it-and-see attitude. This early period of technological innovation turns you and your staff into valuable assets for your community. You have an expertise that has been dearly bought by your institution, and its investment in you and the conversion projects you manage means that you have its attention. Don't waste your chance to help others understand what you have learned about the role digital surrogates can play in the information services of your institution. For whatever reason your institution has embarked upon a digital conversion project, you now have an opportunity to explore how digital technology can help you with your work and also to educate key decision makers about what digitization can and cannot do for students, teachers, researchers, and managers.

7. This point is made forcefully in Peter S. Graham, "New Roles for Special Collections on the Network," *College and Research Libraries* (May 1998): pp. 232–239.

UNDERLYING PRINCIPLES

Guidance Not Guidelines

This book does not lay out best practices or guidelines, but presents a methodology for decision making called digital benchmarking. We advocate this approach because no one right way to do things applies in all circumstances. Guidelines are by their very nature contextual, informed by the assumptions under which they were produced. They may not scale beyond those constraints or to other environments. As Sabinne Süsstrunk, formerly at Corbis, pointed out, "choosing the correct scan resolution is dependent on the purpose of the archive," not on some absolute.[8] Despite the strong temptation to extrapolate from one situation to another, what works for the Library of Congress probably will not work for a local historical society.

Guidance is inductive, providing the means for

8. Sabinne Süsstrunk, "Image Production Systems at Corbis Corporation," *RLG DigiNews*, 2, no. 4 (August 15, 1998), www.rlg.org/preserv/diginews/diginews2-4.html.

What Users Want from Digital Image Collections

Beth Sandore, Coordinator, Digital Imaging Initiative
University of Illinois at Urbana-Champaign Library

What users do with digital images depends on the type of images being used and the desired level of use. Most digital images can represent two basic categories of material: text and graphical objects. The reader of a digitized book needs to be able to view page images in various sizes and resolutions, page sequentially backward and forward, look for specific pages, consult the table of contents and index, and search the full text for occurrences of words, phrases, or concepts. The user of a pictorial image collection may have different needs, such as zooming in to analyze the painting technique used in a watercolor.

General Findings from Recent Studies

▸ *Users tend to be impatient with long waits.* They want to retrieve pages quickly, especially if they are navigating among pages or sections of a document.

▸ *Users need both image and text versions.* Although they value having access to images that act as surrogates of documents, they also want text files to be able to conduct full-text searches.

Screen Display

▸ *Users are generally content with current image quality.* Image-quality ratings in the Museum Educational Site Licensing Project (MESL) depended on the equipment used to display the images, but most of the respondents considered the quality of the image displays to be above average.[1]

   ▸ *Text image size and quality*: Completeness and legibility of pages with minimal scrolling is the primary user requirement for text-based documents. The Internet Library of Early Journals (ILEJ) study notes that users indicated high satisfaction ratings with the clarity and legibility of images that were presented onscreen, which varied from 120 to 200 dpi to accommodate differing page and font sizes at a screen size of 800 × 600 pixels (originally scanned at 400 dpi).[2] Another important factor was limiting scrolling to one dimension.

▸ *Graphical image size and quality*: Users want multiple views to support their different research needs and some are interested in tonal and color fidelity. Analysis of the MESL database indicated that consistency across image quality is a necessity in consortial databases.[3] Informal guidelines for three sizes of images emerged: 1) thumbnails, ranging from 50 to 100 pixels high; 2) screen-size images, averaging 400 pixels high; and 3) full-size images as they were supplied by the museums, sometimes reformatted into minimum compression JPEGs for Web delivery. Recent surveys of the AMICO Library™ of art images have brought mixed reviews of image quality. Some specific criticisms include noticeable loss of detail and poor tonal quality and color appearance.[4]

▸ *Large format graphical images such as maps are difficult to display and fully comprehend online.* When using these collections, users need access to tools that provide them with zoom, pan, and peripheral-view capabilities.

Navigation and Manipulation

▸ *Users prefer a simple interface.* They want to get to the pages that interest them quickly, without getting lost in multilevel menus. Recent interface-design research for the Library of Congress National Digital Library Program has produced tools that are based on sound interface and navigational design principles:

   ▸ Flatten the navigational hierarchy within collections.

   ▸ Minimize scrolling and jumping.

   ▸ Support quick judgment of relevance with previews and overviews of collections.

CONTINUED ON PAGE 5

institutions to specify, create, and implement their own guidelines according to local specifics instead of a predefined set of options. Guidelines are the byproduct of this process, not the starting point. They should be based on the reasons for digitizing, the nature of the source documents, the institutional mission, available resources, the technical infrastructure, and user requirements and capabilities. Guidance begins with understanding the context, provides a process for data gathering and decision making, points to available standards and best practices, offers a means for assessing their applicability, and finally leads to the development of locally appropriate guidelines.

Digital Collections as Institutional Assets

After a decade of experimentation, many institutions are beginning to shift from a project orientation to program development. With this transition comes an appreciation for the value of digital image collections. A movement toward "digital

---

CONTINUED FROM PAGE 4

▸ Provide primary information very early in search interactions.[5]

▸ *Users want to navigate the structural and intellectual content of image collections.* They want to page backward and forward, jump to a specific page, and check the table of contents or index.[6]

▸ *Users want to manipulate graphical images.* At least two studies suggest that graphical image database users need to be able to:

  ▸ Zoom in to view detail.

  ▸ Examine two or more images side by side.

  ▸ Save search result sets.

  ▸ Sort search result sets.

  ▸ Export images into other software.

  ▸ Produce high-quality printouts.

  ▸ Annotate images with comments and save them to a notebook.

  ▸ Use image-editing tools.[7]

Search Functions

▸ *Users need a variety of search functions.* Depending on the image, these can range from SGML or XML marked-up full text to the recognition of picture features.

▸ *Preferences for access points vary according to the collection and the type of image.* In the MESL and AMICO studies, users performed a substantial number of searches for known items, by creator or artist, title of a work, or style or period.[8] Research with other types of picture collections suggests that users pose queries that contain a preponderance of subject terms, both specific and nonspecific.[9]

▸ *Search engines, query formulation, and database indexing affect search results.* The MESL Project demonstrated that queries in databases that contained identical sets of images and text could yield vastly different result sets. The type of search engine and the level of indexing and query construc-tion produced profoundly different search results for the same data set.[10]

▸ *Users want database search options.* Especially where many collections are available, users want to specify a search across multiple databases simultaneously or select one or more databases for a specific search.[11]

Sidebar Notes

1. Beth Sandore and Najmuddin Shaik, "Findings of the Instructor/Student Survey," in *Delivering Digital Images: Cultural Heritage Resources for Education*, The Museum Educational Site Licensing Project, vol. 1, ed. Christie Stephenson and Patricia McClung (Los Angeles: The Getty Information Institute, 1998), p. 113; Geraldine Gay, Robert Rieger, and Amanda Sturgill, "Findings of the MESL Casual User Survey," in *Delivering Digital Images*, p. 125.

2. "Internet Library of Early Journals (January 1996–August 1998) A Project in the eLib Programme: Final Report," (March 1999) www.bodley.ox.ac.uk/ilej/papers/fr1999/fr1999.htm.

3. Howard Besser, "MESL Implementation at the Universities," in *Delivering Digital Images*, pp. 70–85.

4. Michael Robertson and Samantha Powell, "The AMICO Testbed at RIT" (presentation at the Art Museum Image Consortium University Testbed Research Meeting, Pittsburgh, PA, Carnegie Mellon University, June 3–4, 1999).

5. Gary Marchionini et al., "Interfaces and Tools for the Library of Congress National Digital Library Program," *Information Processing and Management* 34, no. 5 (1998): pp. 535–555.

6. Elizabeth J. Shaw and Sarr Blumson, "Making of America: Online Searching and Page Presentation at the University of Michigan," *D-Lib Magazine* (July 1997) www.dlib.org/dlib/july97/america/07shaw.html#page.

7. Sandore and Shaik, "Instructor/Student Survey," p. 119; Samantha K. Hastings, "Evaluation of Image Retrieval Systems: The Role of User Feedback," *Library Trends* 48 (fall 1999) in press.

8. Sandore and Shaik, "Instructor/Student Survey," p. 119; Gay and Rieger, "Findings of the MESL Casual User Survey," p. 3.

9. Linda H. Armitage and Peter G. B. Enser, "Analysis of User Need in Image Archives," *Journal of Information Science* 23, no. 4 (1997): pp. 287–299.

10. Besser, "MESL Implementation," p. 85.

11. Ben Shneiderman, Don Byrd, and W. Bruce Croft, "Clarifying Search: A User-Interface Framework for Text Searches," *D-Lib Magazine* (January 1997) www.dlib.org/dlib/january97/retrieval/01shneiderman.html.

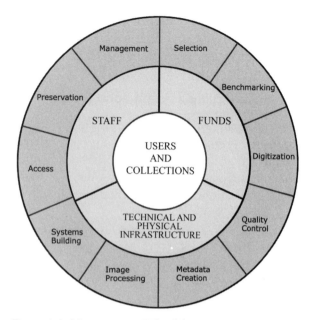

Figure 1.1 Management Wheel for Digital Imaging Programs

ecology" is emerging, replete with arguments for creating digital files that can be repurposed or recycled beyond their original circumstances. The value of digital files over the longer term requires careful resource management, beginning with a recognition that digital images are not just second class surrogates of analog resources but institutional assets in their own right. Institutions must select wisely, invest in proper creation, build and maintain a robust technical infrastructure that includes trained staff, and integrate digital imaging initiatives into their overall organizational goals. This includes developing a strategy for sustainability that is goal driven rather than technology driven, entailing the means for priority setting tied to resource allocation.

We advocate a holistic approach to digital imaging that builds on an appreciation for the interrelation among the processes and resources involved in managing digital image assets over time, from their initial selection and use to their long-term health. Each stage in a digital collection's life affects what comes after. For example, decisions made at the point of selection affect decisions governing digitization, which in turn can affect processing capabilities and preservation strategies. In figure 1.1, we place both collections and users (current and future) at the center of digital program development, as

they provide the touchstones against which to measure all other variables. Radiating out from them are institutional resources, including personnel, finances, and the technical and physical infrastructure. These elements will enhance or constrain digitization programs. The outer circle represents the processes that encompass digital imaging programs. This figure shows the organic nature of digital imaging, with interdependencies connecting goals, resources, and processes.

Social Informatics

Rob Kling of Indiana University defines social informatics as the study of the social aspects of computerization, taking into consideration the interaction within the institutional and cultural contexts. A successful digital imaging program creates collections that different communities use easily and effectively. Kling cautions that:

Good application design ideas are neither obvious nor effective when they are based on technological considerations alone. Their formulation requires understanding how people work and what kind of organizational practices obtain.[9]

One of our guiding principles throughout this book is the need to understand the stakeholders' perspectives. Digital imaging initiatives involve many players, including librarians, archivists, system analysts, programmers, administrators, curators, funders, scholars, and general users. These initiatives will succeed to the extent that they take into consideration the varying needs of stakeholders. The Social Construction of Technology (SCOT) model, which Trevor Pinch of Cornell University articulated in the mid-1980s, is based on the premise that technology is shaped by social factors and is influenced by the norms and values of the social system.[10] To evaluate the Making of America (MOA I) collection, Cornell's Human-Computer Interaction (HCI) Group adopted the

9. Rob Kling, "What Is Social Informatics and Why Does It Matter?" *D-Lib Magazine* 5, no. 1 (January 1999), www.dlib.org/dlib/january99/kling/01kling.html.
10. Trevor Pinch, "The Social Construction of Technology: A Review," in *Technological Change: Methods and Themes in the History of Technology*, ed. Robert Fox (Newark, NJ: Harwood Academic, 1996).

Make Your Digital Imaging Program Successful: Involve the Stakeholders

Don Willis, Manager, Connectex, LLC
Drew Lathin, President, STS Design Consultants, LLC

Integrating digital imaging initiatives into an institution's basic mission can have major consequences that will reverberate throughout the organization. Not only does any technological change appear threatening, but long-term commitment to providing digital access can require redistribution of institutional resources. For this to succeed, it is imperative to involve the stakeholders in the planning process. There are two major reasons for this:

▸ First, to ensure all the application-specific use of the technology is properly evaluated. What to digitize, in what order, at what specifications, how to handle oversize and fragile documents, whether to enhance, and if so to what degree, are just a few of the technical conversion issues to consider. Other decisions affect the value of the collection to your audience: How should the collection be indexed, accessed, and retrieved? How should the output system be configured? What type of software should be used? What capabilities should the display have? Will printing be permitted? It is absolutely critical that those most familiar with how the materials will be used have input on these decisions.

▸ Second, to avoid major upsets in the existing socio-cultural system. For example, new approaches will be required, managers may feel threatened by their employees gaining new skills, and curators of existing collections, who rely on orthodox means for managing resources, could feel that their function will become obsolete. In addition, the dynamics between users and providers of information will change. If the introduction of new technology and the existing socio-cultural system are not aligned, the results will never be optimized. Overt or covert resistance, if not explicitly acknowledged early in the program, could result in slow or partial deployment. The reward system may have to be changed to encourage innovation and the adoption of the new technology.

It is easy to see how the decision to initiate a digital imaging program can affect stakeholders in different ways. It should also be clear that many of the decisions to be made will have economic consequences, so it is critical to get the best information from the subject matter experts and end users early in the design phase. The biggest single source of failures for major technological implementations comes from not involving the stakeholders from the ground up.

The Traditional Way

Until recently, major system-development processes typically involved the "waterfall methodology": development is completed at one level at a time and everything flows downhill. The levels are system requirements, design specification, user approval of the design, system development, system implementation, and turnover. This methodology offers only one opportunity for the stakeholders to provide input, when they approve the design specifications. For many technological implementations, this is either too late or is insufficient to be effective. Technology deployment typically takes a long time and between user approval to turnover things can change any number of ways. The organization changes, users change, requirements change, and technology changes, among other things. If the implementation spans more than six months, there is a good chance that the stakeholders' input may no longer be relevant.

A Better Way

A better approach is called the rapid application development (RAD) methodology. Using this methodology, the introduction of technology proceeds like a spiral. The implementation team makes a first cut at the technical requirements and specifications. After general agreement with the stakeholders, a prototype is built based on the core functionality. The stakeholders then work with the prototype, providing feedback to the developers. The developers then modify the specification and the process goes through several more iterations, finally resulting in a program that is both technically sound and responsive to the needs of the stakeholders.

Parallel with the RAD methodology, stakeholders and users can be surveyed to ascertain their concerns about the technology or approach being taken. A matrix can compare the technical requirements of the initiative to relevant socio-cultural issues to determine potential sources of resistance. Appropriate changes to workgroup structures, reward systems, managerial skill levels, roles and responsibilities of users, and decision-making mechanisms can be tested. Managers can then determine what changes are necessary in the organizational system to facilitate rapid and effective deployment of the new program.

To compare the two approaches, consider sending a rocket to the moon. The waterfall method is like aiming the rocket at the point in space where the moon is at the time of launch; it will surely miss its target. However, if (as in the RAD model) the rocket uses feedback on its position relative to the moon to correct its course, then the rocket will indeed hit its target. The bottom line is that for a major technological transition to be successful, it must involve feedback from the stakeholders it will affect, both those within the cultural institution's staff and those who rely on its resources.

The authors welcome comments and suggestions. Contact them at dwillis@connectex.com and dlathin@stsdesign.com.

SCOT model.[11] The interviews with librarians (including managers, archivists, and preservation specialists) and designers (including software designers, database specialists, and evaluators) involved in the MOA I project revealed that they had different priorities, and interpreted and used terminology differently. Analyzing the manuscript digitization project at the National Library of the Czech Republic, Stanislav Psohlavec argues that researchers and technologists interpret concepts differently:

For example, ask a technician what "frequent access" to data signifies to him. His answer will be in minutes or hours. For a researcher, a manuscript taken from a vault to be studied once a week is highly used.[12]

Successful design and evaluation of digital image collections includes examining and balancing stakeholders' differing notions and priorities. "Cultivating enough mutual understanding to move forward," Michael Ester of Luna Imaging warns, "will demand substantial effort and more than a little inefficiency."[13]

Acknowledging the Larger Environment

Digitization transcends the constraints of physical form and place, allowing for the co-location of dispersed materials as a uniform body of knowledge. Once documents are digitized and made accessible on the Web, they become an integral part of the larger information environment. This highly desirable—and unavoidable—close integration with the external world requires a new approach. As Deanna Marcum, president of the Council on Library and Information Resources, once said, "The notion of a stand-alone digital library seems anathema to its intent."[14] Throughout the last decade, cultural and educational institutions have

started the transition from being guardians and service providers of primarily print collections to balancing services for both print and electronic materials. This transition requires institutions to collaborate closely to ensure efficient use of their resources, including finances, personnel, hardware/software, and space.

The virtues of collaborations at the national and international level cannot be overemphasized. As illustrated throughout this book, collaborations are accomplishing some of the finest work in the areas of digitization, metadata, digital preservation, and image management. However, collaboration comes with its own set of challenges. Educational and cultural institutions are still compelled to justify their existence to their funders by safeguarding their distinctive identity. Collaboration involves synchronizing the missions, cultures, and policies of the participating institutions. Incorporating the differing needs of stakeholders may be an intricate task even in small in-house projects; it is amplified in interinstitutional collaborations, especially international ones.

Collaboration has paid off well in areas where international agreement has been reached on standards for digital imaging. Cultural and educational institutions have always appreciated standards that enable them to share resources. In the digital realm, standards become even more crucial as interoperability greatly enhances the virtues of online, shared information and helps guarantee access well into the future. Throughout this book, we highlight relevant existing and emerging standards and protocols.

At the same time, we argue for maintaining flexibility even as decisions are implemented. Peter Hirtle suggests a work-in-progress attitude to system building. Maintaining flexibility also requires sustained monitoring of technological developments. This is particularly true of technology-based initiatives that depend on their larger environments. For example, legislation and case law are redefining legal rights in the digital realm and will require vigilant attention as the changes will have significant impact on access.

THIS BOOK'S OUTLINE

This book offers the intellectual contributions of more than 50 individuals. Sidebars complement

---

11. Julian Kilker and Geri Gay, "The Social Construction of a Digital Library: A Case Study Examining Implications for Evaluation," *Information Technology and Libraries* 17, no. 2 (1998): pp. 60–69.

12. Stanislav Psohlavec, "Digitization of Manuscripts in the National Library of the Czech Republic," *Microform & Imaging Review* 27, no. 1 (winter 1998): p. 21.

13. Michael Ester, *Digital Image Collections: Issues and Practices* (Washington, DC: Commission on Preservation and Access, 1996), p. 20.

14. Deanna Marcum, comment at a SOLINET conference, "To Scan or Not to Scan: What Are the Questions?" Atlanta, Georgia, May 2, 1996.

Table 1.1. Digital Imaging Introductory Information

▸ Besser, Howard and Jennifer Trant. *Introduction to Imaging.* Santa Monica, CA: The Getty Art History Information Program, 1995. www.getty.edu/gri/standard/introimages/index.html

  Introduces digital imaging technology and vocabulary as they relate to the development of image databases, and outlines the areas in which institutional strategies regarding the use of imaging technologies must be developed.

▸ Kenney, Anne R. and Stephen Chapman. Digital Imaging for Libraries and Archives. Ithaca, NY: Cornell University Library, 1996. (ordering information: www.library.cornell.edu/preservation)

  Provides an introduction to the central issues associated with the use of digital imaging technology in libraries and archives, including a theoretical and technological overview. Advocates a common vocabulary and set of perspectives from conversion to presentation.

▸ "AHDS Publications," *Arts and Humanities Data Service*, ahds.ac.uk/public/pub.html

  Offers several series that address creation, management, and distribution of digital image collections. The Guides to Good Practice series is particularly useful, providing practical instruction in applying standards and good practice to the creation and use of digital resources.

▸ "Digital Toolbox," *Colorado Digitization Project*, coloradodigital.coalliance.org/toolbox.html

  Provides links to general resources, bibliographies, initiatives, and clearinghouses on selection, scanning, quality control, metadata creation, and other project management issues. Also offers a glossary of digital imaging terms.

▸ "Digital Imaging Basics: An On-line Tutorial," *Preservation and Conservation, Cornell University*, www.library.cornell.edu/preservation

  Created by the Department of Preservation and Conservation, Cornell University Library and funded by the National Endowment for Humanities, provides an introduction to digital imaging covering several related issues.

▸ *eLib Supporting Studies*, Preservation Studies, www.ukoln.ac.uk/services/elib/papers/supporting

  Managed by the British Library Research and Innovation Centre, offers several reports on creating and preserving digital image collections. One of the goals is to compare various digital preservation strategies for different data types and formats.

▸ *PADI: Preserving Access to Digital Information*, www.nla.gov.au/padi

  The National Library of Australia's PADI site, offers a subject gateway to digital preservation resources. Includes current information on digital preservation-related events, organizations, policies, strategies, and guidelines. Also includes glossaries of terms that are relevant to digital information.

▸ "PRESERV—The RLG Preservation Program," *RLG*, www.rlg.org/preserv

  Offers supporting materials, such as RLG project reports and the bimonthly *RLG DigiNews* to support institutions in their efforts to preserve and improve access to endangered research materials. The "RLG Tools for Imaging" section includes a worksheet for estimating digital reformatting costs, and guidelines for creating RFPs for digital imaging services.

▸ "Building Image Archives," *TASI (Technical Advisory Service for Images)*, www.tasi.ac.uk/building/building2.html

  Funded by the Joint Information Systems Committee (UK), provides information on creating, storing, and delivering digital image collections. Also lists events and information resources of interest to those involved in digital imaging initiatives.

---

each of the nine chapters with more in-depth coverage of some of the significant issues and projects. As compilers and editors, we wanted to present different perspectives and provide balanced, international coverage of the issues.

In chapter 2, Paula de Stefano discusses criteria for selecting materials for digital conversion. The accompanying sidebars provide testimonies about various selection strategies at a variety of institutions, including special collections, research libraries, consortia, and the Library of Congress.

The remaining chapters cluster into three groups. In chapters 3 and 4, we discuss digital image creation. Chapter 3 advocates digital benchmarking for decision making, which can be applied to both conversion and presentation requirements. We propose a methodology for translating analog

informational content into digital equivalents and argue for creating a richer digital master than needed to meet immediate purposes. Chapter 4 covers how to implement quality control programs and how to assess image attributes, as well as covering the basics of color science.

Chapters 5–7 focus on practical aspects of image management and use. In chapter 5, Carl Lagoze and Sandra Payette raise the challenges and issues to developing metadata schemes for digitized material. They cover several current initiatives that are significant for digital image collections and discuss in detail the functions of metadata, including resource discovery, presentation and navigation, rights management, access control, administration, and preservation. In chapter 6, John Price-Wilkin reminds us that the creation of image

Table 1.2. Continuing Information

Web-Based Publications

▸ *Ariadne*, www.ariadne.ac.uk

Published quarterly by the UK Office for Library and Information Networking (UKOLN), reports on progress and developments within the Electronic Libraries Programme, often covering issues related to digital imaging.

▸ *CLIR (Council on Library and Information Resources) Publications*, www.clir.org/pubs/pubs.html

Frequent reports and research briefs on national and international digital imaging and preservation initiatives.

▸ *Current Cites*, sunsite.berkeley.edu/CurrentCites

Annotated monthly bibliography of selected articles, books, and electronic documents on information technology. While broad in scope, helps keep up with the changes and trends in the digital library realm. To subscribe, send the message "subscribe Cites your name" to listserv@library.berkeley.edu.

▸ *D-Lib Magazine*, www.dlib.org

Monthly, offers articles, commentaries, and briefings that support digital library research. Like Current Cites, covers cutting-edge digital library research and often includes articles related to digital imaging.

▸ *Journal of Electronic Publishing*, www.press.umich.edu/jep

Focuses quarterly on current issues and trends in electronic publishing, covering issues from creation to delivery of electronic information. Many of the issues are also of interest to those involved in digital imaging initiatives. Published by the University of Michigan Press.

▸ *RLG DigiNews*, www.rlg.org/preserv/diginews

Focuses bimonthly on issues of particular interest to those who are involved in digital imaging and digital preservation initiatives. Produced for the Research Libraries Group by the staff of the Department of Preservation and Conservation, Cornell University Library, with partial funding from the Council on Library and Information Resources (CLIR).

Electronic Mailing Lists

These electronic discussion groups frequently announce new digital imaging projects and report on ongoing initiatives. They also include information on digital imaging–related conferences, meetings, and training programs.

▸ Digital Libraries Research Forum (DigLib)

To subscribe, send the message "SUBSCRIBE diglib Your Full Name" to listserv@infoserv.nlc-bnc.ca.

▸ IMAGELIB

To subscribe, send the message "SUB imagelib Your Full Name" to listserv@listserv.arizona.edu.

▸ Digital Librarianship (DigLIBNS)

To subscribe, send the message "SUBSCRIBE diglibns Your Full Name" to listserv@sunsite.berkeley.edu.

▸ PADI Forum

Specifically dedicated to the exchange of news and ideas about digital preservation issues. To subscribe, send the message "SUBSCRIBE padiforum-l Your Full Name" to listproc@nla.gov.au.

---

files is never an end in itself. In addition to discussing systems building, he addresses technical aspects of image processing to create added value in the access system. In chapter 7, Peter Hirtle discusses evaluating and selecting image management systems to organize, manage, and provide access to images and accompanying metadata. Categorizing image management systems, he highlights their advantages and disadvantages.

The last two chapters focus on moving digital initiatives from projects to sustainable programs. Chapter 8 provides a synopsis of the current state of digital preservation from an imaging perspective and suggests a framework for long-term management policies. Chapter 9 presents the case for mainstreaming digital imaging initiatives, offers a management strategy for bringing about the transition from projects to programs, and investigates the financial implications of this shift.

ADDITIONAL RESOURCES

We recognize that to fully benefit from this book requires a basic understanding of the technology. Table 1.1 lists introductory resources. We have limited our discussion of very technical information to forestall making this publication obsolete even before it gets printed. That highly relevant information is available on the Web posed a quandary, since the average lifetime of such documents is estimated at 44 days.[15] We chose to cite online sources, but made an effort to include references with dynamic, updated content and more stability. In addition, table 1.2 lists publications, electronic discussion groups, and Web sites that will help readers keep up with the new information.

15. Brewster Kahle, "Preserving the Internet," *Scientific American* 276, no. 3 (March 1997), www.sciam.com/0397issue/0397kahle.html.

2 Selection for Digital Conversion

Paula de Stefano
The Barbara Goldsmith Curator for Preservation
Head, Preservation Department
New York University Libraries

SELECTION CRITERIA for digital conversion are so varied and overlapping that it is difficult to organize and assess them coherently. There are many legitimate reasons for favoring one selection methodology over another, but the success and efficiency of a project will suffer if the wrong choice is made. Examining all of the issues involved in selecting materials for digital conversion before launching a project will assist in accomplishing both immediate and long-term goals.

Some of the issues that determine the selection of materials for digital conversion include the intellectual content or scholarly value of the materials, available funding opportunities, and the desire to improve access to a collection or to enhance analytic penetration of the texts. Pedagogical utility, whether for classroom use, curriculum support, or distance education, represents another set of motivating factors that may instigate the selection of specific materials. Other motivations may spring from the need to reduce handling and use of fragile or heavily used original materials or from a desire to develop collaborative, resource-sharing partnerships with other institutions, either to create virtual collections and increase access, or capitalize on the economic advantages of a shared approach. In some cases, simply promoting institutional strengths may catalyze a conversion project.

Whatever the catalytic factors behind the decision to convert materials to digital form, the selection process is further refined along a continuum that will require assessment and reassessment in successive stages. For example, whether to convert an entire collection of materials or limit conversion to a logical subgrouping of the collection is a deci-sion that needs to be made early on; likewise, the choice between theme-based or format-based selection. Costs and space savings may play a role in the selection decision, legal and copyright considerations may apply, and technological requirements may demand special kinds of support—the array of colluding circumstances can seem endless.

COPYRIGHT

The influence copyright has on selection is the first crucial issue to address. For good reasons, in *Selecting Research Collections for Digitization*, the authors very pointedly start the discussion under the heading "Copyright: The Place to Begin." They caution that "digital projects must be undertaken with a full understanding of ownership rights, difficult as they often are to obtain, and with full recognition that permissions are essential to convert materials that are not in the public domain."[1] In her sidebar, Georgia Harper suggests that a case for making digital fair-use copies might be made if there is no real alternative for obtaining permission or purchase of electronic versions at fair price. Cornell adopted a "reasonableness" standard for determining copyright status of 20th-century brittle books to be digitized and microfilmed that limited copyright search efforts to the printed Catalog of Copyright Entries.[2]

1. Dan Hazen, Jeffrey Horrell, and Jan Merrill-Oldham, *Selecting Research Collections for Digitization* (Washington, DC: Council on Library and Information Resources, 1998), p. 2, www.clir.org/pubs/reports/hazen/pub74.html.
2. Samuel G. Demas and Jennie L. Brogdon "Determining Copyright Status for Preservation and Access: Defining Reasonable Effort," *Library Resources & Technical Services* 41, no. 4 (October 1998): pp. 323–334.

What About Copyright?

Georgia Harper

Copyright Attorney, Office of General Counsel, University of Texas System

Digital imaging projects always raise copyright questions. Before addressing the most fundamental questions, let's review copyright basics.

▸ Copyright protects original works of authorship fixed in a tangible medium of expression.

▸ The starting point for answering the question, Who owns this work? is always the creator, unless the work is made for hire. Works made for hire are either works created by an employee within the scope of employment or works properly characterized contractually as works made for hire, so long as the parties to the contract follow all the detailed requirements contained in the work made for hire statute. The person or entity employing the creator is considered the creator and owner of works made for hire. Transfers of copyright are called assignments. They must be made in writing and signed by the owner. Otherwise, copyright can be transferred only by a will or intestate succession.

For example, for unpublished letters the author of the letters is the copyright owner. Usually, the author does not transfer copyright to anyone, including the recipient, so the archive that owns the letters does not typically own the copyright for all the correspondence, even if the donor has transferred her rights in the collection. The copyright passes to owner's heirs just like ordinary personal property, with only minor exceptions.

▸ Copyright protection begins and ends at different times depending on when the work was created. The Sonny Bono Copyright Term Extension Act (Pub.L. No. 105-298), passed in October 1998, extended protection so that newer works are protected from the moment of their fixation in a tangible medium until 70 years after the author's death. Older works are protected for 95 years after date of first publication. Works for hire are protected for 95 years after publication or 120 years after creation, whichever comes first. Lolly Gasaway, director of the Law Library and professor of law at UNC, maintains a chart (at www.unc.edu/~unclng/public-d.htm) that summarizes all the different terms of protection for published and unpublished works.

▸ Copyright owners have exclusive rights to do and authorize the following: make copies, create derivatives, and distribute, display, and perform their works publicly.

▸ The rights of users include fair use and performance rights in face-to-face teaching and educational transmissions, among others. Unless authorized by law, users must get their rights from owners.

Copyright and the Digital Image Archive

Answer these questions to decide whether to digitize a document:

1. *Is the work to be digitized in the public domain or does it attempt no more than to faithfully reproduce a work in the public domain?* If either answer is yes, you may digitize the work without securing anyone's permission. Note, however, that if your imaging project results in no more than a faithful reproduction of the original, it is unlikely that your digital files will be protected by copyright, because the reproduction will probably not meet the prerequisite of originality.

2. *If either answer to question 1 is no, was the donor of the material the copyright owner and did the donor give your institution the right to digitize without permission?* Donated original works come with some implied rights, even without a written donor agreement expressly granting broad rights. It may be reasonably assumed that the donor implicitly granted the right to take advantage of new technologies to continue to use the work for the same purposes for which it was donated. Of course, it is better to get rights in writing. Words such as "the right to use the work for any institutional purpose, in any medium" would be helpful additions to a donor form. It is important to note on your donation form whether the donor or seller had copyrights to convey or license.

3. *Is making and providing access to a digital copy considered a fair use?* This is difficult to answer, but several principles make judging fair use easier. If you understand copyright owners' concerns, you can address them in your practices. Electronic copies floating around cyberspace are a copyright owner's nightmare. A university could minimize this threat by allowing access to protected images only to students who are registered for classes, and by changing access codes at the end of each semester. Also, the university should advise students that their rights to use the images are limited to personal, research, study, and other academic purposes. Another major concern for copyright owners is that cultural institutions may not be inclined to purchase digital images if they can make their own.

To protect their interests in developing educational markets, some copyright owners argue that fair use does not cover digital archives; however, this argument is not so persuasive as long as there is no real alternative to making digital fair-use copies. Contrast how easily users can obtain permission to digitize or acquire digital access to journal and textbook materials with the dysfunctional permissions process for images. Where vendors have not made images reasonably available electronically at a fair price, fair use is a viable alternative to seeking permission or purchase. On the other hand, where an image easily may be obtained electronically at a fair price, purchase is the better alternative. The Copyright Crash Course at www.utsystem.edu/ogc/intellectualproperty/cprtindx.htm contains more detailed guidance on fair use in this and other contexts.

The law provides important protection for good-faith judgments about fair use. Even if you and your institution lost a suit over something in your digital archive, if the judge

CONTINUED ON PAGE 13

For the purposes of selection, the question of copyright must be answered at the earliest possible moment. Obtaining copyright permission is not always possible and can derail a project that appears otherwise straightforward. Manuscript materials present some of the most difficult problems, especially in collections that have been donated with implied rights, as opposed to written agreements. To date, many institutions have avoided the issue completely by limiting their selection decisions to collections or subsets of collections that are in the public domain. Others meet it head on, as does the Ad*Access Project at Duke University's Digital Scriptorium. This project gives full copyright information for the images—including a section on the United States Copyright Law, as well as instructions for obtaining permission to use the images and for obtaining reproductions and citation information.[3]

3. "Copyright and Citation Information," *Ad* Access*, scriptorium.lib.duke.edu/adaccess/copyright.html.

CONTINUED FROM PAGE 12

could see that you had a good-faith basis for believing what you did was fair, she could reduce the damages against you to zero. No one wants to be sued, but this protection may discourage potential plaintiffs when you and your institution have a good-faith basis for your actions. To demonstrate good faith, have and follow a reasonable fair-use policy. The University of Texas System's Comprehensive Copyright Policy is at www.utsystem.edu/ogc/intellectualproperty/cprtpol.htm.

If an underlying work is not in the public domain, or the donor did not implicitly or expressly grant rights to digitize without further permission, or you conclude that fair use does not apply, you will need to get permission to digitize the work or acquire access to it by licensing it. Remember, however, that *licenses are negotiable!* Understand your users' needs, and negotiate for sufficient access and rights to accommodate them. Negotiated access will likely replace reliance on fair use for creating digital archives, so negotiating skills will become more valuable over time. The Copyright Crash Course includes information to improve these skills and obtain more favorable license terms.

4. *How can your institution protect its digital assets?* Exact copies of works in the public domain are not protected by copyright law. Also, fair-use copies and those made with the express or implied permission of the owner do not convey copyright in the digital copy. As a result, your digital archive may lack inherent copyright protection unless you own the rights over the original documents. If you do not, you still must guard the copyright owners' interests in protecting developing markets and limiting unauthorized distribution over the Internet.

SELECTION TO ENHANCE ACCESS

Increasing Accessibility

Enhanced access to remote research materials is one of the most—if not *the* most—attractive features of electronic resources. Improved access to a repository's collections is a valuable contribution to a user community and can satisfy both local and national interests. Institutions of higher education fulfill very focused faculty and student needs, while public institutions must satisfy a larger, more diverse population. In some cases, despite the public versus private status of an institution and regardless of the local constituency, many repositories have grasped the opportunity introduced by electronic technology to share their resources with the broadest audience of users—even a global one. As Steve Hensen points out, most rare and unique materials held in special collections must be used onsite, and "anything that can be done to make these materials accessible remotely makes the job of basic historical and literary research that much easier."[4]

For example, in the University of North Carolina at Chapel Hill's Documenting the American South (DAS) project, teachers, students, and researchers all have immediate access to digitized slave narratives, first-person Southern literature, Confederate imprints, and materials related to the church community. DAS provides a vast audience with a powerful single point of access to an extensive breadth of primary documentation.

The widespread appeal of improved access to research materials is extremely attractive. It makes these materials available for simultaneous use by potentially competing users. It also means that collections can be used remotely, eliminating the need to visit a particular repository. This is a great service to researchers and increases the utility of a collection. A physically remote location may thwart access to materials and result in an underutilized collection. When electronic access is provided, use is likely to increase. "Broader access, as it creates

4. Steven L. Hensen, "Primary Sources, Research, and the Internet: *The Digital Scriptorium* at Duke" (paper presented at the annual meeting of the American Historical Association, New York City, January 4, 1997), scriptorium.lib.duke.edu/scriptorium/hensen-aha97.html.

a new community of users, can also facilitate more active scholarship."[5]

The democratic appeal of increased accessibility can be recognized and exploited to benefit a statewide constituency, as in the case of an Illinois initiative. The Illinois Digital Academic Library (IDAL) addresses the problem of providing common access to information for students, faculty, and staff of its member libraries. IDAL serves postsecondary institutions accredited by the Illinois Board of Higher Education and belonging to the Illinois Library and Information Network (ILLINET), approximately 150 academic libraries in the state. Not only does this program help maintain an even level of information access, it offers its participants an opportunity to meld financial as well as digital resources.

The Colorado Digitization Project (CDP) is a similar statewide endeavor "to create an open, distributed, publicly accessible digital library that documents crucial information for the residents of Colorado."[6] CDP has established a collaborative effort that is open to libraries, library systems, archival organizations, selected pilot schools, historical societies, and museums throughout the state. Content, in this case, is limited by the research and information needs of five "user segments" identified by the CDP Advisory Council as casual users, students (K–12 and lifelong users), information seekers and hobbyists, scholars and researchers, and the business community. The project has characterized each of these groups in a table indicating content, design, and retrieval issues to be addressed.[7]

Collaborative efforts between institutions and consortiums are springing up nationally and internationally as well. One such example is the United Kingdom's national electronic resource initiative. As ambitiously described by the Higher Education Funding Council's Joint Information Systems Committee (JISC),

the future development of the distributed, national electronic resource . . . aims to . . . support academic research; postgraduate and undergraduate teaching and learning, especially in the field of open and distance learning; and provide better support for the management of information systems.[8]

Facilitating New Forms of Access and Use

Digitization can extend access by enabling the use of material that is impossible in its analog form. The Digital Atheneum Project at the University of Kentucky, for example, is exploring the use of digital technology to reveal the hidden content of historical documents that have been badly damaged over time.[9]

Beyond data recovery, however, digital conversion projects can focus more narrowly on computer enhancements to enable deeper analysis of a resource than traditional research methods allow. In this way, something more than digital conversion is achieved, introducing added value to the content of a scanned collection. Searchable databases and texts marked up according to standard language formats allow analysis and manipulation of textual content that can lead to new kinds of research tools and discoveries.

Digitally converted texts that have been enhanced by optical character recognition (OCR) can be searched in ways that are otherwise impossible without the aid of a computer. Computer-assisted textual analysis is now used to authenticate texts previously questioned and of dubious origins. By conducting searches in the files of the Making of America project, scholars of the Middle English Dictionary at the University of Michigan pushed back the etymology of a number of words when they discovered their use decades earlier than had been previously documented.

Text encoding is also used to mine the depths of a document, to explore and reinterpret its contents, or for critical analysis or comparative study. Scholarly use of this tool in conjunction with digital image conversion is becoming more widespread

5. Hazen, Horrell, and Merrill-Oldham, *Selecting Research Collections*, p. 5.

6. Collection Policy Working Group, "Collection Policy, June 1999," *Colorado Digitization Project*, coloradodigital.coalliance.org/select.html.

7. "Market Segments and Their Information Needs," *Colorado Digitization Project*, coloradodigital.coalliance.org/users.html.

8. Committee on Electronic Information (CEI)—Content Working Group, "An Integrated Information Environment for Higher Education: Developing the Distributed, National Electronic Resource (DNER)," *JISC (Joint Information Systems Committee)*, www.jisc.ac.uk/cei/dner_colpol.html.

9. W. Brent Seales, James Griffioen, and Kevin Kiernan, "The Digital Atheneum—Restoring Damaged Manuscripts," *RLG DigiNews* 3, no. 6 (December 15, 1999), www.rlg.org/preserv/diginews/diginews3-6.html.

and invaluable to research. The Electronic Text Center at the University of Virginia conducts digital conversion projects that endeavor to create electronic documents endowed with internally searchable features, allowing the researcher to analyze the textual content of literary and historic works. The Center's mission "to create an on-line archive of standards-based texts and images in the humanities," coupled with their commitment to create collections in partnership with subject librarians, special collections curators, local museums, faculty and students "near and far," is a fine example of electronic collection building that goes beyond the physical co-location of analog resources.[10]

SELECTION BASED ON CONTENT

Intellectual content may be the overriding causal factor when deciding to digitize a collection, especially if current research trends and demands conspire to make it a timely topic of inquiry. Of course, intellectual value alone may not be enough of a reason to digitize a collection. For example, if a collection is already being used by the local research community without competition, it may be better to leave the material as is and use limited resources to scan other materials.[11] Ample justification must support any decision to digitally convert a collection, and subject specialists familiar both with the collection and how it is used will be essential navigators of a selection decision when content is the motivating factor.

Virtual Collection Building

Nowhere is broader, electronic access more fully realized than in the collaborative development of virtual collections. The William Blake Archive, for example, was conceived as an

international public resource that would provide unified access to major works of visual and literary art that are highly disparate, widely dispersed and more and more often severely restricted as a result of their value, rarity, and extreme fragility.[12]

Eight American and British institutions contribute to the Blake Archive, as well as one major private collector. Their goal is to "restore historical balance" to the two sides of Blake's art, his visual art and his poetry, "through the syntheses made possible by the electronic medium."[13] This accumulation represents an entirely new resource that promises to enhance scholarly research, providing comprehensive access to materials that would otherwise remain distinct resources. Another example of virtual collection building is the Advanced Papyrological Information System (APIS) project. APIS was initiated through the efforts of papyrologists at a number of American universities "to integrate in a 'virtual' library the holdings from their collections." Under the auspices of the American Society of Papyrologists, a technology committee oversees and coordinates the work, as well as sets methodology and standards for image capture and cataloging.[14]

Critical Mass

The notion of enhanced electronic access strongly implies and, some would argue, requires volume and comprehensiveness in order to make such access worthwhile. Thus, it is claimed that for any digital library to be viable and useful, a critical mass or corpus of research materials must be collected and made accessible to scholars, faculty, students, and the general public.

Participants in the Museum Educational Site Licensing Project (MESL) identified the lack of comprehensiveness as a significant barrier to the widespread use of image databases in teaching:

For most users, even a critical core will not offer a comprehensive corpus. Many faculty teaching with MESL images vocalized a need for a "critical mass" of images that would approach the corpus size of their analog slide libraries.[15]

10. David Seaman, "Goals and Mission," *University of Virginia Library, Electronic Text Center*, etext.lib.virginia.edu/mission. html.

11. Hazen, Horrell, and Merrill-Oldham, *Selecting Research Collections*.

12. *The William Blake Archive*, ed. Morris Eaves, Robert N. Essick, and Joseph Viscome, www.iath.virginia.edu/blake.

13. Ibid.

14. Roger Bagnall, "Advanced Papyrological Information System, A Joint Project of Columbia University, Duke University, Princeton University, The University of California, Berkeley, The University of Michigan, Université Libre de Bruxelles, Yale University: Narrative," odyssey.lib.duke.edu/papyrus/texts/APISgrant. html.

15. Howard Besser, "Digital Image Distribution," *D-Lib Magazine* 5, no. 10 (October 1999), www.dlib.org/dlib/october99/10besser.html.

The prospect of a critical mass of digitized resources is precisely what most excites scholars beginning to use the Internet for their work. Without a critical mass, none of the time savings or convenience inherent in Web-based research can be fully realized. When scholars and faculty speak in terms of content, comprehensiveness seems to be key to satisfying their electronic research needs.

Although a whole host of qualifying factors such as limited funding, staff, and technical expertise may work against the creation of a critical mass, just the opposite can occur. As Ross Cole-

Digitizing the Record of a Colonial Culture: Ferguson 1840–45
Ross Coleman
Collection Management Librarian, University of Sydney

There is a certain circularity about content in any digital project. Initially, content represents the sum and substance of the endeavor. As the project moves into production, content becomes a purely technical objective, dominated by concerns about quality, production, cost, and access. After digitization, the value of the content as subject returns, albeit tempered by its transformation into a new format.

The period 1840–45 was seminal in Australian history, recording the emergence of an independent colonial character. The 75 periodical publications of this era listed in John Ferguson's definitive *Bibliography of Australia* formed the subject of the Australian Cooperative Digitisation Project (ACDP). The collaboration involved three of the largest research libraries in Australia: the National Library of Australia, the State Library of New South Wales, and the University of Sydney Library, the project leader. The decision to digitize comprehensively the contemporary publications of a key period in Australia's political, social, and literary history determined both the importance and the character of the project and dictated many of the subsequent technical processes.

Librarians at the three institutions had discussed microfilming the Ferguson titles for preservation purposes, but the benefits of digitization for access became equally compelling. Leading scholars working in Australian literature and history responded enthusiastically to the proposal for a project involving both microfilm and imaging. Of more consequence to the project, however, was their insistence on taking a comprehensive approach—the value of the project, they argued, was in the full record of the period 1840–45. Collaboration and digitization enabled this comprehensiveness, supporting the merging of distributed physical collections into one complete virtual collection. The project's innovative approach, coupled with broad academic support, gave the proposal sufficient stature to gain funding from the Australian Research Council in 1996.

Microfilming was considered the most effective and least-damaging means of preparing content for digitization in a production process. The microform masters provided a stable long-term preservation format. Pragmatically, microfilming took advantage of substantial institutional and vendor expertise in Australia. Less expertise was available for the kind of digitization proposed in the project, so developing vendor experience in digitizing heritage material was also a fundamental element of the project, and was seen as a long-term benefit. However, due to the inexperience of both project partners and vendors, the decision to outsource led to many months of delay as both par-

ties tested and retested technical specifications, addressed internal and external communication issues, resolved production complications, and reviewed the quality of output.

Project staff faced other challenges. We were unprepared, for instance, for how long it took to compile a complete set of materials from several libraries, identify the best copy for microfilming, and check if titles were already microfilmed and whether the film was suitable for digitization. Having taken a comprehensive approach, we were committed to handling all the physical variations of the publications, which ranged in size from quarto to broadsheet, and the variation within any one title, from foxing and discoloration to the use of varying fonts on the same page. This diversity presented challenges for image quality, file size, and delivery mechanisms. The practical application of specifications that had been based largely on documentation from other projects was also a challenge. We tested these specifications in a pilot phase to provide some practical experience and a basis for expected image quality.

At the end of the production stage, we had completed 65 of the 75 periodical titles, representing about 65,000 images. The other ten titles either could not be located or were only available in microform that did not lend itself to conversion. To enhance access to the journals, OCR is being applied to all internal indexes or contents lists, as well as to the full text of several titles, to investigate providing an uncontrolled keyword search facility. The journals are accessible both through the project site at www.nla.gov.au/ferg and as individually cataloged titles linked through the Web interface of Kinetica, the Australian national bibliographic database.

As the first major digitization project in Australia, we were perhaps overly ambitious imaging a comprehensive record of significant material, while trying to establish a contract production process. Although we are still grappling with some delivery and interface problems, the benefits of the initiative are becoming apparent. The material available so far is being used for study and research around Australia, the vendors have work from an increasing number of clients, successful collaboration has been proven and is continuing, and each partner has also established ongoing digital services.

Selecting a comprehensive, historically viable body of literature has been both the strength and weakness of the ACDP. As an initial collaborative effort, this project strained both staff and resources, but it has been the value of the content—unique access to the contemporary record of a significant historical period—that has underpinned and substantiated the endeavor.

man points out in his sidebar, the idea of comprehensiveness played a big part in defining the content of the Australian Cooperative Digitisation Project (ACDP). Leading academics enthusiastically endorsed the ACDP for its inclusion of the full record of a key period in Australia's political, social, and literary history. At face value, comprehensiveness is a logical approach with understandable, theoretical appeal, as well as historical ties to traditional collection-building methods. But, practically speaking, as Coleman warns, the consequences of a sweeping approach are serious and need to be clearly understood in order to fully prepare for the work involved.

The Applicability of Microfilm Selection Models

Institutions may identify materials for digital conversion according to recognized measures of content. This was common practice, and still is, in the selection process for preservation microfilming. Following the "great collections" model established by the Research Libraries Group, many institutions used RLG's now defunct Conspectus rating to measure a collection's depth in various subject areas.[16] This selection method recognized the stature of research collections built over time and the need to preserve them intact, practically volume by volume. Used in conjunction with microfilm technology, there is ample justification for such comprehensiveness: microfilm technology lends itself well to the massive attempt to preserve great stores of brittle books and is especially well suited for the preservation and storage of low-use materials. With digital conversion, however, access is stressed over other considerations. It makes little sense to scan low-use materials—an inherent and unavoidable flaw in the argument that favors comprehensiveness in digital collection building—and the costs associated with doing so could be seen as irresponsible.

Perhaps a more useful selection approach, also propagated under the influence of microfilm technology but with more appropriate application to digital technology, is Ross Atkinson's insightful solution to preservation treatment decisions. He

divides library materials into three classes: class 1 includes special collection and unique materials, such as rare books and manuscripts; class 2 "consists of high-use items that are currently in demonstrable demand for curriculum and research purposes"; and class 3 are low-use or less frequently used research materials.[17] This model could be easily adapted to focus more appropriately on the two most logical categories of research materials for digital conversion: special and unique materials and materials for which there is considerable demand. The sidebar by Geri Bunker Ingram and Carla Rickerson covers the consequences of selecting special collections materials. Carl Fleischhauer's sidebar explains how educational content was the primary force behind selection for the American Memory project at the Library of Congress.

Scholarly Input

Another traditional selection methodology relies on input from scholars and other researchers. A good example of such an approach is found in the seven-volume *The Core Literature of the Agricultural Sciences*, which involved detailed analysis of the fundamental literature of a whole discipline, based on the input of hundreds of scholars.[18] This compendium informs selection decisions for cooperative reformatting projects (both microfilm and digital) among members of the United States Agricultural Information Network (USAIN).

Columbia University Libraries' Online Books Evaluation Project provides an example of a detailed, methodical approach to the process of enlisting an invested user community of faculty and graduate students in the planning process for a digital library. The purpose of this pilot project was to examine "how scholars interact with various classes of traditional print-on-paper books" as a way of informing future selection decisions for digital conversion. Although the report acknowledged the potential of online books to provide substantial service to faculty and students, and rec-

16. Margaret S. Child, "Selection for Preservation," in *Advances in Preservation and Access*, vol. 1, ed. Barbara Buckner Higginbotham (Westport, CT: Meckler Publishing, 1992).

17. Ross Atkinson, "Selection for Preservation: A Materialistic Approach," *Library Resources & Technical Services* 30 (October/December 1986): pp. 344–48.

18. Wallace C. Olsen, ed., *The Core Literature of the Agricultural Sciences*, 7 vols. (Ithaca, NY: Cornell University Press, 1991–1995).

ommended selecting books with the greatest utility (such as reference works) or broad subject coverage, equal concern was placed on overcoming real and imagined barriers to use posed by the technology. Providing online access without addressing these concerns risked "alienating the academic community."[19]

In a context that extended outside of a single repository to a scholarly community at large, the Perseus Project at Tufts University actively employed editorial boards of scholars to develop the content of digital collections to satisfy pedagogical needs.[20] Proceeding through scholarly dis-

19. Mary Summerfield, "Online Books: What Roles Will They Fill for Users of the Academic Library?" *Columbia University, Digital Library Collections, Online Books Evaluation Project*, www.columbia.edu/cu/libraries/digital/texts/paper.
20. Perseus Project home page, www.perseus.tufts.edu.

Selection of Special Collections Materials for Digitization at the University of Washington Libraries

Geri Bunker Ingram, Coordinator, Digital Initiative Program, University of Washington Libraries
Carla Rickerson, Head, Special Collections, Manuscripts and University Archives
University of Washington Libraries

To understand the impact of digitization projects in special collections at a major research library, one must first consider the rising importance of primary source materials for education at all levels. Although commercially provided information is available to any library with the budget to afford licenses and use fees, it is increasingly the locally owned, primary source materials held in special collections departments that distinguish a library from its peers. These are the "jewels in the crown" of the research library, and to serve their constituencies to the fullest extent, libraries are working to extend online access to unique and rare materials. Thus, the University of Washington (UW) Libraries, as publisher, must mine the treasures buried in the Special Collections, Manuscripts and University Archives Division in order to serve the increasingly sophisticated information needs of its diverse users, from university scholars to innovative eighth-grade history teachers.

Of course, not all primary materials on today's campuses reside in the libraries. Faculty maintain growing collections of teaching materials in digital format and are increasingly seeking out the library as partner in managing these resources. At the University of Washington, the availability of a high-performance image archiving system (CONTENT) and large departmental 35-mm slide collections has provided the opportunity for a completely new library service: creating and hosting visual imagery teaching collections. Such opportunities must be exploited to develop the tools needed to digitize and provide access to the libraries' special collections; these efforts in turn serve the researchers' need to integrate geographically dispersed, multiformat materials. Local development can also gain outside support, such as the award from the Library of Congress/Ameritech National Digital Library Competition, which is partially funding development of the American Indians of the Pacific Northwest, a collaboration of the UW, the Eastern Washington State Historical Society, and the Museum of History and Industry.

Intellectual Description
A well-researched collection with adequate finding aids is a better candidate for digitization than one that needs a lot of preparatory work. Intellectual control is a prerequisite for digital imaging of special collections because they are not self-contained, well-ordered materials. Many institutions lack expertise and time for the research to prepare collections for digitization. This research may range from preparing an inventory for a collection of papers to identifying, dating, and providing the historic context for a collection of photographs. Precious staff resources are required simply to acquire and minimally process such materials. Consequently, the University of Washington concentrates on digitizing photographic collections and finding aids for manuscripts, rather than full-text documents. Even where research time is affordable, integrated library system technologies have by and large ignored archivists and their need for highly structured access tools. Where the staff has constructed finding aids, they often have organized the material on stand-alone and paper-based systems. Selecting a particularly interesting collection often demands that resources be reallocated in order to fully research and describe the items for electronic access.

Conservation
Because special collections materials are rare or unique, a balance must be struck between the risk to the collection and the use to which it is put physically and digitally. Every collection must be evaluated in terms of its value as a primary source and how much handling it can tolerate. A realistic assessment of the need for conservation and for preservation of the collection may determine the scope and depth (and cost in money and time) of a given project.

Implications for Staff and Users
After selecting a collection to digitize, special collections staff quickly learn that users rarely appreciate its scope and relationship to other collections. Ease of access can lead users to wildly unpredictable expectations. Specialists and undergraduates alike might well fall prey to at least three misconceptions: first, the popular myth that "everything is online." In most cases not everything relevant to the inquiry will be available digitally.

CONTINUED ON PAGE 19

cussions, as well as newsgroups, listservs, and open consultations with specialists in the field, the Perseus Project has developed a collection of materials on the archaic and classical Greek world. Staff members regularly consult with advisors from colleges and universities across the country.

In his sidebar, Ross Atkinson suggests that "citations in the digital publications of local scholars" should drive retrospective digitization. In so doing, digitization encourages use by connecting the author to the library and to future readers. Providing access to used collections has the added benefit of satisfying more of the population the repository is trying to serve and thereby does the greatest good.

CONTINUED FROM PAGE 18

Even assuming a well-integrated system and a large percentage of online material, curators are often forced to restrict many items. Sometimes individual aspects, versions, or formats of a collection must be sealed for a fixed period to ensure donor privacy or anonymity.

The assumption of comprehensiveness is usually accompanied by the second misconception: "everything here is true and accurate." The digital arena cannot provide the user with a context for a collection as a reading room can. Because users without sufficient grounding in the relevant literature rely on digital collections of personal papers and photographs, more resources must be allocated to verify captions and describe collections/images. Some context must be provided online to assist remote users in understanding the content of the material.

The importance of this concept cannot be overemphasized. Perhaps the greatest concern weighing on special collections staff is preserving the intellectual integrity of digitized materials. Powerful technologies enable all manner of data manipulation, and "downstream" interference is very difficult to detect and control. In the electronic environment, we must make every effort to educate the remote user about the problems for authenticity and accuracy inherent in the medium.

As a byproduct of collocating dispersed collections, we may also create an increased demand on reference service for our partner institutions. That is, readers who are not under our direct care may expect (or simply need) mediated assistance from the curator with whom we are collaborating across institutions—whether or not that human resource is available.

The third naïve presumption, that "everything online is free," ignores the cost of material acquisition, copyright clearance, and preparation. Our users expect to access traditional materials without fees for service. While the library continues to provide traditional services for free, we must also offer innovative services and products in new formats. The University of Washington has several services for which it charges end users, such as high-quality reproduction of images. Special Collections staff are also seeing a substantive increase in the number of orders for photos and requests for research into the nondigitized portions of digitized collections. It is yet to be seen whether the library can offer—and whether the market will bear the full cost of—reproducing digitized images on demand.

Legal Implications

The issues of intellectual property and privacy rights may determine whether or not a collection can be digitized and made available. New-found interest among widening audiences has spurred efforts to digitize materials previously unavailable except in secure reading rooms. For example, the serious researcher is now joined by members of the high school debate team, each of whom has a legitimate interest in those senatorial papers acquired long ago. All too often, libraries, museums, and archives find that they simply cannot determine whether they have the right to scan and display materials, even though they may own the physical piece. That is, it may be more difficult to determine the copyright status of unpublished materials than published. Personal papers often involve many correspondents—a donor may have given over all of her rights, but that does not cover the rights of individuals who have written to her. In addition, privacy rights and specific donor restrictions often compound the issue. For this reason, it is imperative to negotiate electronic-use rights and maintain a written understanding for materials that may someday be scanned for public digital display.

Implications for the Library as Organization and Repository

Digital projects almost always take more time and resources than expected, and this is particularly true when they involve special collections materials. And preliminary anecdotal evidence shows that success can eventually pose a problem for collections: enhancing access to a subset of a collection through digital reproduction ultimately increases handling of the entire set. But a successful digitization program offers significant advantages for the library. As the materials in its special collections are increasingly recognized for their uniqueness and utility, there is increased use of the collection. Donors sometimes become more intrigued with the possibilities for their collections as they appreciate the exposure that digitization could bring. In some cases, digital access gives new life to collections in endangered or even obsolete formats. For example, the UW has received a large collection of unique painted glass lantern slides depicting life in Japan in the 1930s. These fragile materials will only be enjoyed widely if a responsible program of digitization provides secure access.

The ease of access alone can lead to overarching advances in scholarship. Even partial collections and digitized finding aids provide a sense of what is available, saving the researchers travel expenses and valuable time in finding and examining relevant items. Perhaps the most exciting advance is the potential for the creation of new knowledge. Libraries are just now beginning to understand how digital repositories will affect teaching and research within and across the disciplines.

Selecting Collections and Selecting Technology:
American Memory at the Library of Congress
Carl Fleischhauer, Technical Coordinator
National Digital Library Program, Library of Congress

At the Library of Congress, American Memory collections are selected to serve the broad educational community, from grade school through higher education and life-long learning. The Library of Congress digitization program provides access to American historical collections that would otherwise be unavailable and emphasizes special collections of manuscripts, photographs, maps, sound recordings, music, and motion pictures, as well as hard-to-find printed matter. Materials are selected to provide important and interesting content that challenges students' critical thinking skills and leads them back to books. As collections are chosen, the Library receives guidance from advisory committees and expert consultants, and from numerous meetings and workshops for educators.

The desire to serve education and present hard-to-find items has led to the selection of materials with high intrinsic value. For example, the project has digitized the world-famous photographs produced by the Farm Security Administration, rare 19th-century imprints containing first-person narratives from a variety of American locales, lithographic bird's-eye views of American small towns, and Edison's first motion pictures. The Washington and Jefferson papers are obvious treasures of the Library of Congress, while the other selections represent the institution's treasury of vernacular expression. Some American Memory presentations provide access to archival collections in their entirety; others are compilations of materials selected from various parts of the Library.

Value to the field of education is the most important criterion for selection, but other factors also influence our decisions. The program is intended to be a catalyst within the Library, drawing staff into the digitization effort and educating them about its effects and ramifications. Thus, some selections have taken advantage of the enthusiasm of curators or custodians for a particular body of work. The digitizing effort has dovetailed with other Library activities such as exhibitions or special preservation or cataloging projects. For example, the Daniel A. P. Murray collection of African American pamphlets (memory.loc.gov/ammem/aap/aaphome.html) needed conservation treatment and the individual works had never been separately cataloged. Their inclusion in American Memory moved in synch with their deacidification, rebinding, and cataloging. From the start, the digitization effort has received critical support from donors and, when it came time to select which presidential papers to digitize first, a gift from Reuters America, Inc. and the Reuter Foundation led the Library to begin with George Washington and Thomas Jefferson.

The value of the original items has influenced the program's choice of technology. For us, it is more important to protect the originals than to create perfect reproductions. In the case of the presidential papers, for example, we scan existing microfilms to avoid handling the original documents as well as to reduce costs. Although a scan from a microfilm could not match the quality of a scan from the original paper, the Library's film-scanning con-

tractor has done a masterful job, cropping document images on a frame-to-frame basis and enhancing images to provide the best legibility. In fact, image quality has been no more problematic than file transfer rates and storage.

The desire to protect originals also influences how we digitize books. Our paper-scanning contractor developed custom cradles to scan bound volumes using one of the face-up book-scanning devices in relatively widespread use today. Although this device can produce hundreds of bitonal images per day, it is technically less sophisticated than one would wish and the resulting images are perfectly legible but not clean, crisp, and uniform. Since we do not intend to replicate books by printing fresh copies, however, this outcome serves our purposes very well.

Our program produces searchable texts as well as page images for most printed matter and some manuscripts. At first, the American Memory pilot project did not even produce page images, trusting that careful rekeying would result in excellent transcriptions. By 1992, after two collections had been produced this way, we saw that page images were needed to represent the look and feel of the original and to permit readers to verify the transcriptions. Our online presentation of book texts continues to evolve. Recent collections, like *Pioneering the Upper Midwest* (memory.loc.gov/ammem/umhtml/umhome.html) feature page-image sets much more prominently than earlier collections did.

Although our printed-matter and manuscript selections forced us to balance the competing goals of good-quality reproduction, protection of originals, and cost-effective production, other formats have given us great opportunities for uncompromised, high-quality digitization. Our recent photographic scanning projects, for example, have produced digital files capable of handling every job traditionally carried out by conventional photographic copies; for example, providing the imagery that a publisher needs to make a printing plate or that an exhibit program needs to make an enlargement for framing.

These examples illustrate selection driving technology. However, for maps newly available technology drove selection. By the end of the American Memory pilot project, practical ways to scan and present large map images had emerged. Using a large flatbed scanner, the cartographic team makes high-resolution color images in which every line and word is legible. The Geography and Map Division also obtained special software that permits the effective presentation of maps on the Web.

Although at the Library of Congress selection adheres to a central goal of serving the field of education, the process supports other goals as well. Each project has its own identity and chooses the most appropriate technology for its circumstances. It is a tribute to the flexibility of the digital environment and the creativity of the National Digital Library team that these varied collections have been brought together as a coherent offering, a rich corpus ready for exploration and discovery.

SELECTION FOR PRESERVATION

Whether manuscripts, books, photographs, or audio/visual materials, use of library and archive materials contributes to their eventual demise. If materials are old and fragile to begin with, contin-ued handling could cause irrevocable loss of both the artifact and its intellectual content. One way to retain and protect the original is to create a surrogate copy that researchers can handle and use instead. Fairly common practice in the past, both microfilm and photocopy technology have been

---

Selection for Digitization in Academic Research Libraries: One Way to Go

Ross Atkinson
Deputy University Librarian, Cornell University Library

Next to collecting publications, copying publications is probably the oldest continuing responsibility of libraries. The purpose of such copying (be it manual, photographic, or digital) is to enhance access—either across space (making multiple copies for current users) or across time (preserving copies for future users). In deciding what to digitize, therefore, the core question remains the same: Which items in our individual collections warrant such enhanced access?

At least six principles drive the development of most library collections:

▸ *The priority of utility*: Usefulness is the *ultima ratio* for all collection decisions. The greater the potential for an item to be useful, the greater the need to add it to the collection—or to make it accessible online. Predicting utility, however, is notoriously problematic and all research libraries add many materials to their collections that will likely never be used. Resources spent on materials that will never be used are wasted.

▸ *The local imperative*: There is no consciously built national collection. There are only local collections, built with local funding, in support of local needs. Any expenditure of local resources must have a demonstrably local benefit.

▸ *The preference for novelty*: In general, old stuff cannot compete with new stuff. Progress in scholarship depends upon our opting in most instances to add something new rather than save something old. Although historical collections are essential for research, only limited resources can be devoted to the collection and maintenance of older material.

▸ *The implication of intertextuality*: The collection is a set. To add an item to a collection is to create a relationship between it and the other items in the collection. Building a collection therefore always creates new textual relationships.

▸ *The scarcity of resources*: All collection development decisions are fundamentally economic, balancing scarce resources: funding, staff time, shelf space, disk space, and user time and attention.

▸ *The commitment to the transition*: More and more information will become available in digital form. Because this transition will benefit education and scholarship, libraries are responsible for promoting it and assisting users in adjusting to it.

Because of the preference for novelty and the scarcity of resources, we cannot retrospectively convert all of our traditional collections to digital form. We must select a relatively limited subset of our holdings for digitization. Provided that all research libraries undertake such a program, the sharing of digitized resources will not be a violation of the local imperative, because the benefits any one institution receives from access to the digital collections of other institutions should exceed the local costs of sharing digitized material, along the same model as shared cataloging. Whatever the local institution selects for digitization, however, must have a direct and demonstrable local utility. It must also have some form of internal coherence to adhere to the implication of intertextuality; that is, the items selected must have a clear relationship. Funding must be set aside from the library budget specifically for the digitization of library resources. Funding cannot be left in subject budget lines, or it will end up being used for other purposes, because of the priority of utility and the preference for novelty.

How should selectors decide upon which materials to digitize, using this separately allocated funding? In general, excepting for core publications, they should not. For other materials, academic research libraries should allow citations in the digital publications of local scholars to drive selection for digitization. A scholar who intends to publish something in digital form should contact the library in advance, and the library should make every effort to ensure that as many of the cited items as possible will be digitized, so that they can be linked to the citations in the publication. This will require the library to take technical action and, in many cases, negotiate the copyright. This method will fully satisfy the priority of utility, because these materials will have been used by the author and will be used by anyone reading the publication. Because the citations are part of a current publication, they support, at least indirectly, the preference for novelty. Digitization of cited publications also complies with the local imperative, because it directly benefits local scholars. Being cited in the same publication provides the digitized items with a clear context, or intertextuality.

If all research libraries took this action and if each library assumed responsibility for archiving and making accessible what it has digitized over time, a set of retrospectively digitized materials would closely complement "born-digital" publications and form an integral part of the international digital collection. It would also demonstrate to scholars that libraries will provide them with important services not only as readers but also as writers.

used to reproduce books and manuscripts, while image and sound transfers have been made to duplicate audio and visual materials. Today, digital technology provides another method to relieve the use of the original. Selection of materials for this purpose must anticipate the need for high-quality surrogates to satisfy a range of uses. A poorly scanned and reproduced original that does not satisfy the needs of researchers will drive them back to the original.

Unlike the production of surrogate copies, replacement of fragile or damaged materials is the goal when the original is physically unusable and an existing copy cannot be found. In this case, the original may not be retained due to its poor condition. An important tenet of this reformatting process is the preservation quality of the reproduction and adherence to standards that ensure the physical stability and longevity of the reproduction.

Digital technology has not yet been accepted by the preservation community as a stable, long-lived format. Some institutions have used it to produce preservation-quality analog versions of digital materials. Perhaps more than any other institution, Cornell University Library has pioneered the use of digital imaging for preservation reformatting. In 1990, the Department of Preservation and Conservation began experimenting with a prototype scanner and the DocuTech printer, both from Xerox, to create paper replacements for deteriorating brittle books. The paper copies, printed on acid-free paper, met or exceeded the quality obtained using conventional preservation photocopy methods. From 1994 to 1996, Cornell investigated the production of computer output microfilm (COM) from digital image files, concluding that under tightly controlled conditions the COM could meet national standards for image quality and permanence.[21]

Nonetheless several problems have been encountered in using digital technology for preservation reformatting. First, these studies revealed that using digitization solely to create paper or film versions of source documents may not be cost-effective when compared to either photocopying or microfilming. It was justifiable, however, when enhanced access capabilities were also considered. Motivated to serve the twin goals of preservation and access, Cornell has focused much of its digital reformatting efforts on endangered but important 19th-century brittle materials, creating high-quality image files and either paper replacements or COM. User response to both the digital images and paper facsimiles has been favorable.

The second problem is common to other reformatting options: disposition of the original volumes. As with photocopying, the quality of the digital images has been predicated on copying single sheets, which requires disbinding the volumes and trimming the inner margin of the pages. For various classes of materials—even brittle items—some faculty members have rejected the notion of destroying the original in order to preserve it. In the Making of America project, for example, a number of the 19th-century journal runs had to be boxed and sent to the rare books vault for storage, despite the fact that Cornell owns duplicate sets of these titles and paper replacements had been created.

The third problem has been the most problematic. If enhancing access justifies creating digital images, then preservation concerns now extend to the digital files themselves. As discussed in chapter 8, digital preservation is much more complicated than simply providing proper environmental controls for microfilm.

Existing Selection Criteria and Guidelines for Digitization

A number of institutions have developed guidelines for selecting materials for digital conversion. These guidelines often double as policy statements and serve to outline an institution's intentions and digital collection-building strategies. One is Columbia University Libraries' Selection Criteria for Digital Imaging Projects. While not prescriptive, the criteria are meant to be "applied to the process of selecting materials for digital imaging before any new project is initiated."[22] The University of Cali-

21. Anne R. Kenney and Lynne K. Personius, *Joint Study in Digital Preservation, Report: Phase I, Digital Capture, Paper Facsimiles, and Network Access* (Washington, DC: Commission on Preservation and Access, 1992), www.clir.org/pubs/reports/joint/index.html; Anne R. Kenney, *Digital to Microfilm Conversion: A Demonstration Project, 1994–1996, Final Report* (Ithaca, NY: Cornell University Library, 1997); available from "Publications," *Cornell University Library, Preservation & Conservation,* www.library.cornell.edu/preservation/pub.htm.

22. "Selection Criteria for Digital Imaging Projects," *Columbia University, Digital Library Collections,* www.columbia.edu/cu/libraries/digital/criteria.htm.

fornia at Santa Barbara has developed a similar set of guidelines.[23] Although worded differently, both address comparable issues, such as the value of the materials to the scholarly community, whether a digital copy will add value to the existing paper or analog copy, and preservation, funding, and technical support.

Several guidelines furnish a decision-making matrix to assist in selection, while others simply offer a worksheet to accompany the selection guidelines.[24] Janet Gertz's "Selection Guidelines for Preservation" derives selection criteria from tradi-

tional preservation microfilming models, but adapts them nicely for use in digital conversion projects.[25] The National Digital Library Program (NDLP) has a checklist of steps used in the production process for historical collections at the Library of Congress, including selecting collections for digitization.[26] Lastly, the Archivschule Marburg published a list of criteria for digital conversion that resembles a treatise more than a set of guidelines.[27] This is a very detailed analysis of all aspects of a digital conversion project and a thorough examination of such topics as "different aspects of uniqueness in archival collections, need of criteria for establishing intrinsic value [and] procedures for establishing intrinsic value." All of these guidelines and checklists provide useful examples for institutionally specific or project-related selection criteria.

23. "University of California Selection Criteria for Digitization," *InfoSurf: The USCB Library*, www.library.ucsb.edu/ucpag/digselec.html.

24. "Selection for Digitization: A Decision-Making Matrix," *Harvard University Libraries, Preservation*, preserve.harvard.edu/resources/digitization/selection.html; Stuart D. Lee, "Scoping the Future of the University of Oxford's Digital Library Collections," appendix B (September 1999), www.bodley.ox.ac.uk/scoping/report.html. Also see Paul Ayris, "Guidance for Selecting Materials for Digitisation" (paper presented at the Joint RLG and NPO Preservation Conference, Guidelines for Digital Imaging, September 28–30, 1998), *RLG*, www.rlg.org/preserv/joint/ayris.html and Selection Task Force of the Electronic Preservation Committee, National Agricultural Library, "Selection Criteria and Guidelines: Selecting Materials for Digital Preservation at the National Agricultural Library, November 1, 1995," *National Agricultural Library*, preserve.nal.usda.gov:8300/projects/criteria.htm.

25. Janet Gertz, "Selection Guidelines for Preservation" (paper presented at the Joint RLG and NPO Preservation Conference, Guidelines for Digital Imaging, September 28–30, 1998), www.rlg.org/preserv/joint/gertz.html.

26. Library of Congress, National Digital Library Program, "NDLP Project Planning Checklist," *Library of Congress, American Memory*, memory.loc.gov/ammem/prjplan.html.

27. Angelika Menne-Haritz and Nils Brübach, "The Intrinsic Value of Archive and Library Material," *Archivschule Marburg, Digitaler Texte Nr. 5*, www.uni-marburg.de/archivschule/intrinsengl.html.

3 Digital Benchmarking for Conversion and Access

Anne R. Kenney

FEW STANDARDS govern the creation and use of digital images, and in a world of multiple stakeholders and multiple perspectives it may be difficult to agree on a uniform approach that suits all circumstances. Determining how to digitize and present library and archival materials involves a fairly complex decision-making process that takes into consideration a range of issues, beginning with the nature of the source document but encompassing user needs, institutional goals and resources, and technological capabilities. These all map together as a matrix for making informed decisions rather than exacting standards.

WHAT IS DIGITAL BENCHMARKING?

Benchmarking is primarily a management tool, designed to lead to informed decision making about a range of choices and an understanding of the consequences of such decisions. Although actual practice must confirm or modify your decisions, benchmarking can reduce both experimentation and the temptation to overstate or understate requirements. Benchmarking allows you to scale knowledgeably, making effective overall decisions rather than developing them item by item or setting requirements that may apply only to a subset of materials. Benchmarking can facilitate negotiations with vendors for services and products. Having benchmarked your requirements, you can focus the discussion on your institution's needs rather than trying to adapt to a particular vendor's offerings.

Benchmarking supports careful resource management. If you know your requirements up front, you can develop a realistic budget. You can avoid buying the wrong equipment or having to manage image files your institution's technical infrastructure does not support. Perhaps most important, you can avoid promising more than you can deliver. For instance, if your institution is scanning its map collection, you can be realistic about what you can deliver to users' desktops.

At the heart of this approach are:

- *Requirements definition*: Assess your source documents and identify the variables associated with quality, cost, and/or performance. This leads to specifications, such as legibility of the smallest text, preserving color appearance, or fast delivery.

- *Measurement*: Gather good data, characterize your material objectively, and determine the interrelationship of variables to desired outcomes. Measurements include document attributes such as detail size and tonal range, scanning requirements (resolution, bit-depth), and the corresponding imaging metrics (MTF and signal-to-noise ratio).

- *Tolerance values*: Determine how much variance you'll tolerate, as your specifications may not be achievable in a production environment or under all circumstances. For instance, you might allow a vendor to reduce resolution at a certain file size or ensure accuracy of color values within a certain numeric range. Your tolerance may be quite low for some requirements and high for others.

▸ *Verification*: Approve or modify your benchmarks through carefully structured testing and evaluation.

Creating Rich Digital Masters and Derivatives

Although there is no consensus on the appropriate method for determining requirements, there is growing support for creating digital masters that are rich enough to be useful over time and cost-effective. This position presumes that conversion standards are set higher than the immediate requirements. Preservation is one of the main arguments for rich digital masters. Digital files can be created to replace or reduce use of a deteriorating or vulnerable original, provided the digital surrogate offers an accurate and trusted representation. Preservation of the digital files themselves is also served when digital images are captured consistently and well documented. Michael Lesk and others have noted the good economics of converting once, or at least only once a generation, and producing a sufficiently high-level image so as to avoid the expense of reconverting when technology requires or can use a richer digital file.[1] This point is particularly compelling since identifying, preparing, inspecting, and indexing digital information far exceeds scanning costs. Further, Peter Galassi, chief curator of photography at the Museum of Modern Art (MOMA), suggests creating a high-end digital master that is "purpose blind."[2] The archival master can be used for derivatives that meet a variety of current and future users' needs. The quality, utility, and expense of derivatives for publication, image display, and computer processing are directly affected by the quality of the initial scan.

More recently a number of digitization initiatives have stressed the need to develop cultural heritage resources that are not only reusable and sustainable, but comparable and interoperable across disciplines, user groups, and institutional types.[3]

Adopting a consistent approach facilitates integration between "born-digital" collections and the digitized files that institutions create from their retrospective holdings. Michael Ester was one of the first to argue for the creation of rich digital masters (which he called "archival images") and to distinguish them from access images ("derivative images"). An archival image, he argued, "has a very straightforward purpose: safeguarding the long-term value of images and the investment in acquiring them." He saw derivative images as a natural byproduct of the archival images, conforming to current technological capabilities and meeting current user needs.[4]

Factors Competing with Informational Content

Digital conversion begins with defining the informational content of source documents, but other factors affect the ultimate choices for capturing and presenting information. Many projects have had to temper the desire for the best possible image capture with other considerations. Institutions may reach radically different conclusions about image quality for similar documents. One institution may decide to serve the originals to users and set imaging requirements lower, while another may choose higher quality so that it can retire the documents to safer storage or dispose of them.

At times, your choice comes down to time and money: convert fewer items at a high level or more items at reduced quality. Neil Beagrie and Daniel Greenstein of the Arts and Humanities Data Service (AHDS) in the UK summed up the conflict succinctly:

The standards and best practices which promised to facilitate and reduce the cost of a data resource's long-term preservation were not always those which promised best to facilitate and reduce the cost of its intended use. The standards and best practices which promised to ensure a data resource's maximum fitness for purpose were also not always affordable or technically achievable.[5]

1. Michael Lesk, *Image Formats for Preservation and Access: A Report of the Technology Assessment Advisory Committee to the Commission on Preservation and Access* (Washington, DC: Commission on Preservation and Access, 1990).

2. "The Modern('s) Digital Archive," *Photo District News* (October 1998).

3. See, for example, "Why Does the Cultural Community Need Best Practices?" *National Initiative for a Networked Cultural Heritage (NINCH)*, www.ninch.org/PROJECTS/practice/why.html.

4. Michael Ester, *Digital Image Collections: Issues and Practices* (Washington, DC: Commission on Preservation and Access, 1996), pp. 11–12.

5. Neil Beagrie and Daniel Greenstein, "A Strategic Policy Framework for Creating and Preserving Digital Collections" (July 14, 1998) *Arts and Humanities Data Service*, ahds.ac.uk/manage/framework.htm.

Digitizing Science: JSTOR Faces New Challenges

Elizabeth Z. Bennett
Production Coordinator, JSTOR

JSTOR, a nonprofit organization digitizing back issues of scholarly journals, released its first titles to participating institutions in January 1997. By the summer of 1998, JSTOR had digitized more than two million pages, primarily titles in the humanities and social sciences. These pages were digitized as 600-dpi TIFF files, with additional 200-dpi JPEG files created for all grayscale and color illustrations. When we began work on a cluster of general science journals, we realized that their content differs in significant ways from the earlier titles.[1] The scientific journals are heavily illustrated with images ranging from engravings and lithographs through halftones and computer graphics. The importance of these detailed illustrations raised concerns that a 200-dpi JPEG might not adequately capture the information they contain.

In evaluating our options, we considered the needs of our end users, our responsibility to archive the substantive content of each title, the technologies available for creating and viewing images, and the cost of digitizing, storing, and delivering the images. Our decision-making process involved definition of the issues, in-house evaluation of various approaches to digitizing these illustrations, consultation with the Cornell University Library Department of Preservation, discussions with our board of trustees, and negotiations with our scanning contractor.

Careful review revealed that the nature and quality of the illustrations varied significantly from journal to journal and across time. We also evaluated the informational content of the illustrations. Although some are essential to the content of the article, other graphics are purely decorative. Test scans suggested that our standard 200-dpi JPEG might not adequately capture finely detailed 19th-century illustrations and contemporary halftones. We examined factors such as resolution, bit-depth, compression, image enhancement, scanner calibration, and file format. We also discussed the ways in which these images might be used, including onscreen viewing, color printing, and image processing. Our consultants helped us evaluate our findings and the technical options for digitizing this material.

Armed with new questions and with price information from our scanning vendor, we formed a committee to reevaluate the material. The committee included representatives of all the functional groups within JSTOR, including production, technology, and user services, as well as three librarians and our senior programmer. The chair, Ira Fuchs, vice president for Computing and Information Technology at Princeton University, is also a member of JSTOR's board of trustees and JSTOR's chief scientific advisor. We made additional test scans, incorporating suggestions from consultants and the initial study, and compared multitonal JPEG images at 200, 300, and 400 dpi, descreened, sharpened, and enhanced in other ways.

We found essence, detail, and structure as defined in the Illustrated Book Study to be very helpful in determining con-

CONTINUED ON PAGE 27

Factors competing with informational content include:

▸ *Conservation*: The physical safety of the originals may drive digitization decisions, especially if high-quality reproduction threatens the document.

▸ *Economics*: The more resolution, bit-depth, or enhancement, the slower the production and the higher the cost of capture and storage. Some institutions have traded fidelity for greater production or lower conversion costs, especially when they have retained the original materials to serve the most demanding users. In assessing the economy of lowering conversion requirements, consider the full range of your costs—higher capture is more expensive initially, but you may recover costs downstream in image processing or file longevity.

▸ *Technology*: Some institutions have based their conversion requirements on the prevailing technical constraints of monitors, printers, and networks, only to discover that their image files do not lend themselves to reuse as the technology improves. Others have argued for the highest capture possible offered by the technology, but find that they are not able to use those files effectively in their current form. For instance, experts advocate greater than 8-bit/channel capture to represent the full dynamic range of photographs. Some institutions that embraced this requirement have discovered that the technology for capturing, storing, and representing the added bit-depth is not fully available. In the Internet Library of Early Journals (ILEJ) project, another technological requirement—the accuracy of subsequent optical character recognition (OCR) processing—became the most demanding criterion of image quality.[6]

▸ *User needs*: Users' perspectives and requirements can affect conversion and should drive access requirements. At Yale, the Beinecke Library's chief goal in converting the Boswell

6. "Internet Library of Early Journals (January 1996–August 1998) A Project in the eLib Programme: Final Report" (March 1999), www.bodley.ox.ac.uk/ilej/papers/fr1999; NARA also chose 300 dpi for textual documents to be compatible with OCR software, see Steven Puglia and Barry Roginski, "NARA Guidelines for Digitizing Archival Materials for Electronic Access," (January 1998) *National Archives and Records Administration*, www.nara.gov/nara/vision/eap/eapspec.html.

CONTINUED FROM PAGE 26

version requirements in the context of JSTOR's mission to create a reliable electronic journal archive.[2] Essence represents the level of detail that can be seen by the unaided eye at a normal reading distance. In nearly all cases, journal illustrations do not by themselves merit close study; rather, they accompany or enhance points made in the text. We determined that capturing the essence is appropriate for digitizing the important informational content in most journal illustrations. During our assessment, we concluded that a 200-dpi, grayscale JPEG would adequately capture the essence of most of the 300,000 illustrations we estimate will be digitized as part of the general science cluster. The informational content of many of these illustrations does not justify, in our view, the added cost of creating and storing a 300- or 400-dpi image.

The committee did find that some illustrations—estimated at 10% of the total—contained such fine detail that a 200-dpi JPEG did not fully represent the substantive content. Our production process enables us to flag those illustrations that require special treatment. For example, color illustrations with lines finer than $1/100$ inch or with type with an x-height less than 1 mm will be flagged for scanning at 300 dpi with 24-bit color.[3]

We also addressed the related issue of page context for illustrations. Our practice had been to present both an image derived from a 600-dpi, 1-bit TIFF for the entire page and a separate 200-dpi, 8-bit JPEG for each illustration. This JPEG did not include associated captions. We concluded that having all the information about an image in a single file would better serve users. For each illustrated page, therefore, the database will contain a full-page, bitonal TIFF image and a separate JPEG image for each graphic, including the caption and any associated labels, scales, or other relevant text.

Evaluating imaging technologies for the illustrations in the general science cluster was instructive for everyone involved and raised a number of strategic issues of broad importance to

a large-scale project like JSTOR. One major question illuminated by our evaluation is whether a large, evolving project can or should pursue and maintain comprehensive internal technical consistency within the entire database. On the one hand, we do not want to embark on retrospective reconversion of the large number of JPEGs already in the database, especially when those images clearly serve our mission well. On the other hand, we want to remain open to the possibility that new material might require a different approach to digitization and that new technologies might offer additional functionality to our users, either today or in the future. We must be prepared to adjust our approach, but balancing consistency and adaptability requires flexibility that a single digitizing standard cannot offer. In addition, because the material in JSTOR varies widely, choosing a single approach would lead to digitizing many images at lower-than-appropriate or higher-than-necessary levels. Consequently, we have chosen to match the technology to the task, so we can fulfill our mission while ensuring the best use of resources. The experience we have gained during this review has helped us to refine our strategy for continuing to provide an electronic archive that, while rooted in the past, can grow and evolve to keep pace with a dynamic future.

Notes

1. *Science, Proceedings of the National Academy of Sciences, Philosophical Transactions,* and *Proceedings of the Royal Society of London.*

2. Anne R. Kenney and Louis H. Sharpe II, "Illustrated Book Study: Digital Conversion Requirements of Printed Illustrations" (July 1999), *Library of Congress, Preservation,* lcweb.loc.gov/preserv/rt/illbk/ibs.htm.

3. In defining this level of detail, we followed the method outlined in Anne R. Kenney and Oya Y. Rieger, *Using Kodak Photo CD Technology for Preservation and Access* (Ithaca, NY: Cornell University Library, 1998), www.library.cornell.edu/preservation/pub.htm.

papers was to produce a usable product for the Boswell editorial project. Scanning in color provided a more aesthetically pleasing image, but the research benefits of color images could not justify the higher cost and file sizes.[7]

BENCHMARKING FOR CONVERSION

Translating Between Analog and Digital: A Proposed Methodology

Conversion requirements depend on consistent, quantified information about source documents.

7. Nicole Bouché, *Digitization for Scholarly Use: The Boswell Papers Project at the Beinecke Rare Book and Manuscript Library* (Washington, DC: Council on Library and Information Resources, 1999), p. 7; www.clir.org/pubs/reports/pub81-bouche/pub81text.html

Physical dimensions, level of detail, tonal range, and color representation can all be characterized numerically and matched up with digital equivalencies. The process requires a shared vocabulary and standardized evaluation of system performance and digital output. Currently, work is underway on quality metrics and software programs to characterize scanning system performance objectively. A *complete* process that is consistent for all documents, scanning environments, and output devices is still lacking, because our knowledge is unevenly developed and the tools are not fully available.

Consider the necessary components of such a process:

1. Critical information in the source document is characterized in terms of digital equivalencies.

This involves linking perception to numbers, formulas, and systems by:

- translating measurements from the original into digital counterparts, and/or
- identifying test targets that are representative of those key properties, which have been assigned digital values or measure performance, and
- determining additional information such as lighting conditions, luminance values, or correction for human perception.

2. Tolerance or deviation values are used to pass or fail digital products.

3. A digital capture system is calibrated and tested to determine its objective performance and its ability to create digital images that approximate the original documents.

4. Digital images are inspected visually against the originals, including the test targets, and/or analyzed with software. The quality inspection should be replicable or verifiable by other parties.

5. Information on the metrics, process, settings, and requirements are recorded for use in presentation and digital preservation.

This approach may well become commonplace in the future and should lead to smoother vendor–client relationships, better production, improved quality control, and hardware and software that facilitate the process. The work of the Photographic and Imaging Manufacturers Association (PIMA)/IT10 Technical Committee on Electronic Still Picture Imaging to develop standards, test objects, and software is particularly encouraging.[8] Targets (and increasingly accompanying software) link the analog and digital worlds in a standardized fashion and provide the means to detect change along the digitization chain. Leading cultural institutions, most notably the Library of Congress and the National Archives and Records Administration (NARA), have pioneered the use of metric requirements in their requests for proposals for imaging services.[9] Although not totally successful, they represent a critical shift in the definition of requirements and accountability for digital products.

Original versus Intermediate

Whether to use the original or an intermediate for digitization is a matter of debate. Using the original, many argue, provides a better scan and less-complicated workflow. IBM and the Vatican Library Project discovered that color errors added by the film degraded color representation (for an example of color degradation, see plate 1). MOMA opted for direct digital capture of their photographic holdings, citing time savings, better and more immediate control for adjusting color and tonal quality, quicker access to converted images, and the first-generation quality of subsequent copies from a digital file. The California Heritage Collection project found that a photo intermediate introduced complexity and more image transformations into the imaging workflow. In the ILEJ project, staff considered images scanned from microfilm poorer than those created directly from the original volumes, and the University of Michigan experienced a 50% decline in OCR throughput using images scanned from film.[10]

Others have made good cases for relying on a film intermediate. Kevin Donovan points out that film is good for storage, backup, and printing.[11] Excellent results have been achieved by projects such as the Blake Archive, which has scanned color transparencies. In a collaborative project involving papyri, it was discovered that direct scanning could not produce the necessary quality cost-effectively. The project is producing 4" × 5" color transparencies, arguing that these intermediates can serve as the source for digital retakes as the tech-

8. *PIMA/IT10: Electronic Still Picture Imaging*, www.pima.net/standards/it10a.htm.

9. Library of Congress, Contracts and Logistics Services, "RFP97-9 Table of Contents," memory.loc.gov/ammem/prpsal9/toca.html, link to Section C.4.

10. Linda Serenson Colet, Kate Keller, and Erik Landsberg, "Digitizing Photographic Collections: A Case Study at the Museum of Modern Art, NY" (paper presented at the Electronic Imaging and the Visual Arts Conference, Paris, September 2, 1997), p. 3; University of California, Berkeley, The Bancroft Library, "Digitizing the Collection: Overview," *California Heritage Collection*, sunsite.berkeley.edu/CalHeritage/digital.html; "ILEJ Final Report"; John Price-Wilkin, e-mail to Don Waters, Anne R. Kenney, and David Seaman, March 9, 1999.

11. Kevin Donovan, "Anatomy of an Imaging Project: A Primer for Museums, Libraries, Archives and Other Visual Collections," *Spectra, A Publication of the Museum Computer Network* 23, no. 2 (winter 1995–6): pp. 19–22.

nology improves.[12] Experts offer cautions about using intermediates. James Reilly of the Image Permanence Institute suggests that when you create an intermediate from an original film source, do not go down more than one film size or you will see perceptible loss.[13] Others have made recommendations for creating photography in anticipation of digital conversion.[14]

Determining Informational Content: Objective versus Subjective Approaches

Determining what kind and level of information represented in the original source is essential to its meaning and must be conveyed by the digital image file involves both objective and subjective evaluation. James Reilly and Franziska Frey have suggested using the physical properties of the original, human visual power, or a combination of both.[15]

Defining Informational Content Based on Physical Properties
You can peg conversion requirements to the process used to create the original document. For example, for photographs, film resolution can be measured by the size of the silver grains suspended in an emulsion, whose distinct characteristics are appreciated only under a microscope. The resolution depends on film type, size and speed, the chemical processing, and the recording capability of the system used to produce the photograph. Ultimately, this can represent a huge amount of information—the equivalent of 12.5 million pixels can be produced on fine-grain 35-mm film, for

Fig. 3.1 Quality Degradation: Direct Scan (left); Microfilm Scan (right)

example.[16] For graphics, the telltale structural evidence of the process used to create them might be the benchmark. Must a digital image accurately depict the black lace of an aquatint that only becomes visible at a magnification above 25X? Significant information may well be present at that level for some materials and uses, such as medical or biological imaging. But in other cases, attempting to capture all possible information will far exceed the informational properties of the image as distinct from the medium and process used to create it. Consider, for instance, a badly blurred 4" × 5" color negative. The negative may be incredibly information dense—in terms of dynamic range, color, and recording capability—but it does not convey significant information. Obviously, for many practical applications, basing conversion on the physical properties of the originals could overwhelm a digital initiative with recording, computing, and storage requirements.

Defining Informational Content Based on Visual Perception
You can also determine information content subjectively by considering human visual perception. Although offset printing uses 1,200 dpi, most individuals would not be able to discern the difference between a 600-dpi and a 1,200-dpi digital image of a page. Choosing the higher resolution, you add more pixels and increase the file size with little or no appreciable gain. How do you determine an adequate resolution? It is more difficult to define a stan-

12. Roger Bagnall, "Advanced Papyrological Information System, A Joint Project of Columbia University, Duke University, Princeton University, The University of California, Berkeley, The University of Michigan, Université Libre de Bruxelles, Yale University: Narrative," odyssey.lib.duke.edu/papyrus/texts/APISgrant.html.

13. Anne R. Kenney and Stephen Chapman, *Digital Imaging for Libraries and Archives* (Ithaca, NY: Cornell University Library, 1996), p. 154.

14. Ester, *Digital Image Collections*, pp. 8–10; "TASI Photographic Guidelines," *TASI*, www.tasi.ac.uk/building/photoguide1.html.

15. James M. Reilly and Franziska S. Frey, "Recommendations for the Evaluation of Digital Images Produced from Photographic, Microphotographic, and Various Paper Formats" (May 1996), *Library of Congress, American Memory*, memory.loc.gov/ammem/ipirpt.html; and Franziska S. Frey and James M. Reilly, *Digital Imaging for Photographic Collections: Foundations for Technical Standards* (Rochester, NY: Image Permanence Institute, Rochester Institute of Technology, 1999), www.rit.edu/~661www1/sub_pages/page17.pdf.

16. Reilly and Frey, "Evaluation of Digital Images," and Frey and Reilly, *Digital Imaging for Photographic Collections*.

dard quality reference for using visual perception rather than physical properties to determine informational content. Some institutions compare the prints from digital images to the original documents or to analog reproductions. MOMA, for example, has defined imaging requirements for photographs to support 8"× 10" color prints for publication.[17]

For other institutions, the desire to protect or even replace an original has led to a digital quality requirement comparable to that obtained by conventional reformatting. In a series of brittle book digitization projects at Cornell University, printed pages from digital images were compared to the quality of preservation photocopies, and computer output microfilm was judged against national standards for preservation microfilming.[18] The Library of Congress has also determined that digital reproductions can rival analog reproductions as preservation copies for some materials, including manuscripts, printed matter, sound recordings, and possibly photographs up to a certain size.[19] Steve Puglia of NARA agrees that digital technology can create preservation replacements for photographic originals, but he distinguishes between reproduction quality and preservation quality that is based on the information carrying capacity of the original negative or transparency.[20]

Subjective evaluation becomes more problematic when the goal is legibility rather than fidelity. For text-based material, the degree of legibility might be defined as ranging from marginal quality to high fidelity. For nontext-based information, legibility becomes even more difficult to judge. For example, is a fine detail that is just perceptible in the digital image file considered legible? For halftone illustrations, legibility might be defined as being free of visible moiré.

This subjectivity leads to dramatically different quality requirements at institutions imaging the same type of material. Most institutions making sheet music available over the Web have based conversion requirements on legibility, but their conclusions about how to create files differ. Duke University chose 150-dpi, 24-bit color for its collection. The Library of Congress scanned microfilm versions of sheet music as 400-dpi, 1-bit images to accurately capture the music notations. And at the University of California at Berkeley, every printed page of sheet music was scanned at 400 dpi, in color when necessary.[21] Similar differences also characterize projects focusing on maps.[22]

If legibility is a difficult benchmark to standardize, other requirements, such as the tones and colors, diverge even more. Most institutions rely on visual assessments, with little to no appreciation for factors affecting color and tone in the digital realm. Quality requirements are also very context-sensitive. In the Museum Educational Site Licensing Project (MESL) for instance, participants differed in their requirements. University representatives defined high resolution for online display, while the museum partners wanted to produce publication-quality, four-color prints—the file size required to support printing was 20 times larger than that for display.[23]

If you choose visual perception for image-quality benchmarking, also consider the level of view to support. Are the original materials typically viewed from a distance, as posters are, or under normal reading conditions, as textual documents are? Are they scrutinized close up or under magnification, as are papyri? Are they enlarged for printing or projection? The digital images produced in the MESL Project were considered acceptable for

17. Colet, Keller, and Landsberg, "Digitizing Photographic Collections," pp. 5–6.

18. Anne R. Kenney and Lynne K. Personius, *Joint Study in Digital Preservation, Final Report* (Washington, DC: Commission on Preservation and Access, 1992); Anne R. Kenney, *Digital to Microfilm Conversion: A Demonstration Project, 1994–1996, Final Report* (Ithaca, NY: Cornell University Library, 1997); available from "Publications," *Cornell University Library Preservation & Conservation*, www.library.cornell.edu/preservation/pub.htm.

19. Carl Fleischhauer, "Digital Historical Collections: Types, Elements, and Construction" (August 21, 1996), memory.loc.gov/ammem/elements.html.

20. Steven Puglia, "Creating Permanent and Durable Information, Physical Media and Storage Standards," *CRM* 2 (1999) tps.cr.nps.gov/crm/archive/22-2/22-02-10.pdf.

21. "Creation of the Images and Database," *Historic American Sheet Music, Rare Book, Manuscript and Special Collections Library, Duke University*, scriptorium.lib.duke.edu/sheetmusic/techinfo-scanning.html; "Music for the Nation—Digitizing the Collection," *Library of Congress, American Memory, Music for the Nation*, memory.loc.gov/ammem/smhtml/smtech.html; Mary Kay Duggan, "19th-Century California Sheet Music," www.sims.berkeley.edu/~mkduggan/project.html.

22. David Yehling Allen, "Creating and Distributing High Resolution Cartographic Images," *RLG DigiNews* 2, no. 4 (August 15, 1998), www.rlg.org/preserv/diginews/diginews2-4.html.

23. *Delivering Digital Images: Cultural Heritage Resources for Education, The Museum Educational Site Licensing Project*, vol. 1, ed. Christie Stephenson and Patricia McClung (Los Angeles: The Getty Information Institute, 1998), p. 58.

onscreen viewing, but the Columbia University faculty found the quality of the digital image sorely lacking when projected alongside a conventional 35-mm slide of the same object.[24] The most challenging level of view to be supported in using the original—via magnification, projection, or printing to paper or film—could become the quality reference in determining digital resolution requirements based on visual perception.[25]

Aligning Digital Requirements to Document Attributes

The process of matching document attributes to digital equivalencies begins with a careful assessment of which aspects of the original will affect the scanning approach and which must be represented in the digital surrogate. The second step is to determine whether you can represent those attributes by some objective measurement. To assess document attributes, survey and characterize the range of materials to be scanned. How many pages will be converted? What percentage contains color? What is the range of physical page dimensions? Select representative samples of the document types, and their physical condition, as well as of the more challenging items.

Document characterization can be highly scientific, involving specialized equipment, or more subjective, based on physical examination. Densitometers, colorimeters, and spectrophotometers measure reflectance and color values. Other document assessment tools include:

- Transparent ruler with markings to $1/16$ inch, at least 15 inches long

- Eye loupe for examining and measuring characters and other details (10X with metric reticle, measuring to .1 mm)

- Microscope with 30–60X magnification, light source, and reticle measuring to $1/1000$ inch or .01 mm for examining and measuring strokes and very fine features

24. Ibid., p. 72.

25. Michael Ester suggests that institutions base digital conversion requirements on a quality to support the "functional range" of an institution's reproduction source. He argues for using the digital image to regenerate the initial photographic format (print or film) and then comparing the resulting quality against that source and any of its intended uses. Ester, *Digital Image Collections*, p. 11.

- Halftone screen ruler to determine the screen frequency of halftone illustrations and to assess digital print quality

- Viewing booth with fluorescent light source (5,000 kelvin) for viewing reflective originals, or table lamp with controlled light source

- Light table with fluorescent light source (5,000 kelvin) for examining transmissive media

- 18% neutral gray card or 11-step gray wedge for inspecting a photo for subtle color casts

To assess document attributes that affect scanning, consider:

- Physical type, size, and presentation

- Physical condition

- Document classification

- Medium and support

- Tonal representation

- Color appearance

- Detail

Physical Type, Size, and Presentation

Begin by considering the dimensions and physical presentation of representative items. Divide source materials first into those that are reflective—recorded on an opaque support such as paper—and those that are transmissive—recorded on film, glass, or some other transparent base. For transmissive items, note their types (roll film, slides, transparencies) and formats (35 mm, 4" × 5", etc.). If a transparency is to be scanned in lieu of the original, consider the physical dimensions of the original item—or at least the reduction ratio used—as well as the film size.

Reflective originals can be either bound or single leaf. If bound, determine whether you can disbind the items for scanning. Measure the dimensions of the page. Decide whether to capture the full page, including any borders, or whether the digital image should extend beyond the edge of the page to represent its shape. If a page is oddly shaped, imagine laying a rectangle over it. What are the dimensions of that rectangle? Decide whether to include a ruler or scale in the digital image and allow space for it. Consider also the placement of color and/or grayscale bars. Measure height and width in inches; Cornell measures to the nearest $1/16$ inch.

The physical type, size, and presentation of documents also determine the kind of scanning equipment to use and the costs of imaging. For instance, common flatbed scanners cannot accommodate oversize items (usually greater than 8½" × 14") and most bound volumes.[26] Appendix A to the RLG Worksheet for Estimating Digital Reformatting Costs provides a way to match materials to scanning devices based on their physical attributes and scanning requirements.[27] In addition to product literature from various companies, good sources on scanning devices are available on the Web.[28]

The physical size and shape of a document can also have a pronounced impact on resolution and file size. As the dimensions of a document increase, either the file size increases or the resolution decreases. Consider, for instance, two photographic prints, one 4" × 5" and the other 8" × 10". If both are imaged in grayscale at the same scanning resolution (say 300 dpi), the file size for the 8" × 10" is four times larger (7.2 MB as compared to 1.8 MB). If, on the other hand, the file size for each image is the same (say 1.8 MB), then the resolution of the 8" × 10" is half that of the 4" × 5" (e.g., 150 dpi as compared to 300 dpi).

$$\text{File size (in bytes)} = \frac{\text{height} \times \text{width} \times \text{bit-depth} \times \text{dpi}^2}{8 \text{ bits per byte}}$$

26. The University of Toronto reports good success in scanning bound volumes that are not too thick or too tightly bound on the Xerox XDOD flatbed scanner. Karen Turko, letter to Abby Smith, March 19, 1999.

27. "RLG Tools for Digital Imaging," *RLG*, www.rlg.org/preserv/RLGtools.html.

28. See Wayne Fulton, "A Few Scanning Tips," www.scantips.com, for links to manufacturers, buyer's guides, and product reviews. "Digital Camera Reviews," *Digital Photography Reviews*, www.inconference.com/digicam/camera.html, and Nicholas Hellmuth, *Digital Photography*, www.digital-photography.org/default. html, offer reviews of digital cameras and equipment. The Digital Eyes home page, image-acquire.com, is an independent source for information on scanners, digital cameras, and software. The Technical Advisory Service for Images (TASI) Web site includes good basic information on scanners, digital cameras, software, and interfaces: "Image Capture: Hardware and Software Summary," *TASI*, www.tasi.ac.uk/building/image_capture_hw_sw.html. The Center for Digital Imaging, Inc. evaluated many digital cameras under $15,000 and some scanners using a standardized image: "Tests," *Center for Digital Imaging, Inc.*, www.cdiny.com/html/tests.html.

This tradeoff of resolution versus file size has prompted some institutions to constrain imaging requirements for oversized documents or let the resolution be determined by a preset file size. For instance, the National Archives scans maps and drawings up to 11" × 17" (or 187 square inches) at 300 dpi; larger maps and drawings are scanned at 200 dpi.

Digital cameras have set pixel dimensions (e.g., 2,000 × 3,000 pixels). Theoretically, a digital camera can capture documents of any size. As size increases, the distance between the camera and the document increases, as do the size of and distance between each pixel, and the resolution goes down. For instance, the effective dpi for a 4" × 5" document is ten times greater that of a 40" × 50" document captured with the same pixel dimensions. This competition between document size and level of detail is most problematic for documents that are large in relation to the level of detail they contain, such as oversized, highly detailed maps. Other large documents, such as posters, may only contain gross features, so the competition for pixel coverage is less problematic.[29]

The physical size of a document can be critical to its meaning. For example, the scale on a map may make sense only if the reader knows the physical dimensions of the original. Printed versions of digital images often preserve dimensional fidelity, but onscreen images cannot unless the resolution of the image equals that of the monitor. Several projects include rulers or scales within the image or record physical dimensions in the accompanying metadata. In his graduate work at the University of Michigan, Mark Handel developed an applet that lays a "measuring tape" on the image.[30] The Blake Archive developed a Java™ applet, the ImageSizer, which enables users to view Blake's work onscreen at its actual physical dimensions.[31]

29. Anne R. Kenney and Oya Y. Rieger, Using *Kodak Photo CD Technology for Preservation and Access* (Ithaca, NY: Cornell University Library, 1998); available from "Publications," *Cornell University Library Preservation & Conservation*, www.library.cornell.edu/preservation/pub.htm; Kenney and Chapman, *Digital Imaging for Libraries*, pp. 9–12.

30. Mark Handel, "Issues of Scale and Size in Visual Databases," sunsite.Berkeley.edu/Imaging/Databases/Fall95papers/handel.html; [Mark Handel], "Proposed Interface for Scale Information," sunsite.Berkeley.edu/Imaging/Databases/Fall95papers/handel2.html; [Mark Handel], "Java!" www-personal.si.umich.edu/~handel/java.

31. *The William Blake Archive*, ed. Morris Eaves, Robert N. Essick, and Joseph Viscome, www.iath.virginia.edu/blake.

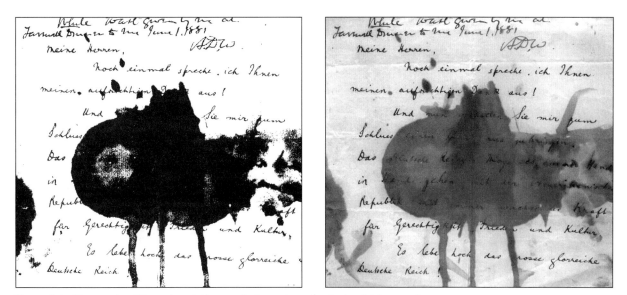

Figure 3.2 Comparing Bitonal and Grayscale Scanning of a Stained Manuscript

Physical Condition

To assess documents, characterize any potential harm to the originals, such as strain on bindings, light sensitivity, brittle paper, or fragile glass plates. Institutions have protected originals by sacrificing image quality requirements, requiring onsite imaging by vendors, training scanning technicians to handle material properly, and coupling conservation treatment and scanning. Several institutions have developed protective cradles and copy stands and have controlled lighting and other environmental conditions to minimize mechanical, radiation, and heat damage. Still others have used film intermediates in lieu of the originals. Several papers and reports describe this work.[32]

Even when the original does not require special handling or equipment, its physical condition may well affect the scanning requirements. Assess, for example, darkening pages, fading ink or burn-through, uneven printing, bleed-through, extraneous markings, smearing, foxing, staining, buckling, and fold lines. Determine whether this information should be captured or minimized in the resulting digital file. Such features often require that the digital image provide tonal distinctions in order to preserve their appearance and/or to distinguish them from the textual information beneath. For instance, a cleanly printed text page may be a good candidate for bitonal scanning; one that is stained or heavily annotated or has darkened requires grayscale or color scanning (see figure 3.2). As a rule, bitonal scanning is cheaper than grayscale and color scanning, but it may become more expensive if additional time is needed to adjust the threshold for each image to ensure adequate representation of tonal information. The centuries-old documents scanned by the Archivo General de Indias were in various states of deterioration, exhibiting water spots, stains, faded ink, and bleed-through. Staff determined that grayscale capture was necessary to convey the informational content and required less operator skill.[33]

If you use a film intermediate, consider the quality and condition of both the original and the film, including scratches, fingerprints, color fading, or

32. Library of Congress, National Digital Library Program and the Conservation Division, "Conservation Implications of Digitization Projects," and "Appendix I, Handout," from "Background Papers and Technical Information," *Library of Congress, American Memory,* memory.loc.gov/ammem/ftpfiles.html; John McIntyre, "Protecting the Physical Form" (paper presented at the Joint RLG and NPO Preservation Conference, September 28–30, 1998), *RLG,* www.rlg.org/preserv/joint/mcintyre.html; Stephen Chapman, "Guidelines for Image Capture" (paper presented at the Joint RLG and NPO Preservation Conference, September 28–30, 1998), *RLG,* www.rlg.org/preserv/joint/chapman.html; Timothy Vitale, "Light Levels Used in Modern Flatbed Scanners," *RLG DigiNews* 2, no. 5 (October 15, 1998), www.rlg.org/preserv/diginews/diginews2-5.html.

33. Pedro González, *Computerization of the Archivo General de Indias: Strategies and Results* (Washington, DC: Council on Library and Information Resources, 1998), www.clir.org/pubs/reports/gonzalez/contents.html.

under exposure. Scanning can amplify existing physical flaws.

Closely related to physical condition is nontextual evidence conveyed by source documents. Several imaging initiatives have asserted the need to represent the look and feel of originals to preserve their historical essence. At the University of Virginia, staff distinguish between preservation imaging, which is primarily concerned with the reproduction of textual information, and archival imaging, which is intended to preserve the sense of the page as an artifact or physical object.[34] In a thoughtful essay on reformatting, German archivists and librarians have argued that the testimonial evidence or nontextual clues that support or clarify the content through external formal features must be considered as part of the document's authenticity or meaning.[35]

This requirement obviously has implications for imaging as well as for metadata. Capturing watermarks, plate marks, paper texture, chain lines, shadows, or the curvature of a page requires grayscale or color capture, and the binding technique, materials, and dimensions have to be described.[36] In addition, this requirement almost certainly increases the time and cost of capture. Some have questioned producing fewer images at higher cost for this level of capture. Referring to manuscript materials, Stanislav Psohlavec of the National Library of the Czech Republic argues that digital imaging should not try to serve the few users who are interested in the manuscript as a physical object. It is difficult to satisfy their needs,

most require access to the originals anyway, and the funds for facsimile imaging could more usefully be used to conserve the originals. "Using digitization to take the place of a microscope is not economical," Psohlavec argues, "and the practice can hardly be defended. Using digitization to get a 'good look' at a manuscript is possible and useful, even today."[37]

Document Classification

Paper and film-based documents may be classified into one of the following categories:

- *Printed text/simple line art*—distinct-edge-based representations that are cleanly produced, with no significant tonal variation. For such documents, rendering detail may be the key goal. Assuming that the digital file conveys only textual information, this material lends itself to bitonal digitization with sufficient resolution. At lower resolution, the scanning device may detect fine features but not fully resolve them. Bitonal scanners capture information in grayscale and then reduce it to black or white pixels. Fine features not fully resolved are interpreted as gray information and may drop out if the threshold is set at a point where their gray value is reduced to white (see figure 3.3). On the other hand, if the threshold is set so that gray values are interpreted as black, fine lines are thickened and other features may fill in. In effect, reducing gray information to black or white reduces image quality dramatically when features are finer than the resolution used to capture them. Grayscale imaging can represent features that are not fully resolved. Unfortunately, grayscale may be more expensive and slower. Carefully assess the detail in printed documents to determine resolution requirements, especially if you plan to use bitonal scanning.

- *Manuscripts*—soft-edge-based representations of textual information that are produced by hand or machine, but do not exhibit the distinct

34. University of Virginia Library, "Producing Digital Images," *The Electronic Archive of Early American Fiction: Comparing Usage and Costs between Electronic Texts and Original Print Editions of Rare Early American Fiction*, www.lib.virginia.edu/speccol/mellon/image.html; Robert Thibadeau and Evan Benoit, "Antique Books," *D-Lib Magazine* (September 1997), www.dlib.org/dlib/september97/thibadeau/09thibadeau.html.

35. Angelika Menne-Haritz and Nils Brübach, "The Intrinsic Value of Archive and Library Material," *Archivschule Marburg, Digitaler Texte Nr. 5*, www.uni-marburg.de/archivschule/intrinsengl.html.

36. Even high-quality images cannot provide all the clues needed for positive identification of some illustration processes. Print historians may rely on the type of paper used, an examination of the surface of the paper to determine the layers, the date of publication, or even the name of the illustrator to make a positive identification. Anne R. Kenney and Louis H. Sharpe II, "Illustrated Book Study: Digital Conversion Requirements of Printed Illustrations" (July 1999), *Library of Congress, Preservation*, lcweb.loc.gov/preserv/rt/illbk/ibs.htm.

37. Stanislav Psohlavec, "Digitization of Manuscripts in the National Library of the Czech Republic," *Microform & Imaging Review* 27, no. 1 (winter 1998): pp. 22–23; Adolf Knoll, "For Whom and How We Digitize," *Web World, Memory of the World, Memoriae Mundi Series, Bohemica 1998*, www.unesco.org/webworld/mdm/czech_digitization/doc/users.htm#For whom to.

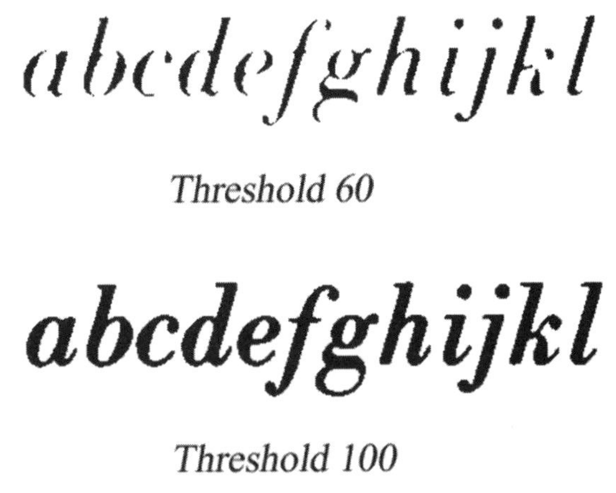

abcdefghijkl

Threshold 60

abcdefghijkl

Threshold 100

Figure 3.3 Effects of Threshold Setting

edges of cleanly printed text. This category also includes graphic materials, such as hand-drawn illustrations, produced with the same writing devices. For such items, detail rendering may be key, but how they were created affects how they are converted. Manuscripts may convey a range of tonal and color information or contain modifications to the initial draft, such as annotations or cross-outs. Most institutions have chosen to scan manuscripts in grayscale or color to maintain the softer-edged features characteristic of handwriting or impact printing (e.g., typewriting). The National Archives scanned all the tex-

tual documents chosen for the Electronic Access Project (EAP) in grayscale because so many of them—for example, typed carbon copies and thermofax—had inherently low contrast.

▸ *Halftone or halftone-like items*—created by a pattern of variably sized, regularly spaced dots, lines, or markings, often placed at an angle. Beginning in the 1880s, halftone printing was used extensively for photographs and artwork in publications. Halftones are particularly difficult to capture digitally, as the screen of the halftone and the grid of the digital image often conflict, resulting in distorted images with moiré patterns (see figure 3.4). Although not strictly halftones, some illustrations such as etchings and engravings also exhibit regular patterns that may result in moiré. In "Digital Formats for Content Reproductions," Carl Fleischhauer describes four approaches to imaging halftone documents.[38]

One of the most consistent ways to scan halftones is to use grayscale capture at a sufficient resolution to minimize or eliminate moiré patterns. A conservative approach would be to scan at four times the screen ruling—the num-

38. Carl Fleischhauer, "Digital Formats for Content Reproductions" (July 13, 1998), memory.loc.gov/ammem/formats.html; David Blatner and Steve Roth, *Real World Scanning and Halftones* (Berkeley, CA: Peachpit Press, 1993), pp. 75–77.

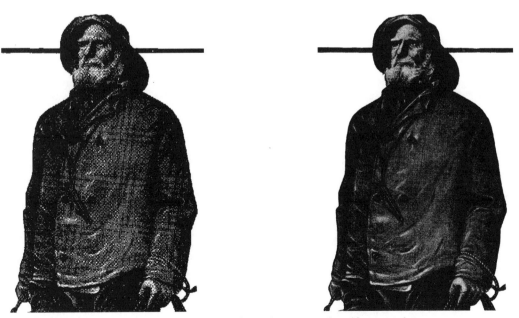

Figure 3.4 Comparing a Halftone Showing Moiré (left) with a Normal Halftone (right)

Benchmarking Digitizing Systems for Photographs

Franziska S. Frey
Research Scientist, Image Permanence Institute
Rochester Institute of Technology

Benchmarking starts with a thorough knowledge of the materials selected for digitization. What is the photograph's type, what is its format, and—in the case of old photographs—what is its state of deterioration? Equally important, who will use the digital files and why? These last questions often cannot be answered today, since future use may not yet have been determined. However, if the files are produced in as open and versatile a format as possible, they can be repurposed for future uses.

You can make decisions objectively or subjectively. For reproducibility and coherence, do both. Carefully define and characterize the important aspects of the object to be scanned. In addition, clearly state how this characterization applies to the proposed imaging systems. A well-designed decision-making process is mandatory for a successful project. It allows planners to see the possible shortcomings of every choice.

Parameters for imaging technologies have only recently been quantified. Despite the available theoretical knowledge and understanding of the different parameters involved, targets and tools to measure these parameters objectively are still missing. Furthermore, open systems, which are used for most digital imaging projects, include modules from different manufacturers. Since the different components influence each other, the overall performance of a system cannot be predicted from any one manufacturer's specifications. An additional hurdle is that more processing is handled automatically in software, making limited information available to the operator.

Once you decide on your parameters, benchmark them. System benchmarking is a twofold process. First, benchmark the imaging system itself to determine its capabilities. In many cases, system capabilities will far exceed your needs. As a second step, benchmark images to be digitized. This will show whether image files meet the requirements for the intended use. Standards are currently being developed for benchmarking scanning systems. Benchmarking systems will help to compare different hardware, giving better information than the manufacturers provide, and should lead to a better understanding of the whole imaging process.

Look at four main parameters when assessing the visual appearance of an image: spatial resolution, tone reproduction, noise, and color reproduction. You can measure and benchmark these parameters with special targets and software, but visual quality can be affected by other choices as well. For instance, spatial resolution alone may be insufficient to produce images as sharp as the original photograph, tone reproduction is affected by all the components of an imaging system and is often difficult to control, and color reproduction includes decisions about pictorial-rendering intent as well as color space and bit-depth. Consider also how visual benchmarks are affected by image processing—sharpening, filtering, noise reduction, and compression. System evaluation should encompass subjective visual assessment as well, which requires that trained evaluators look at images with a basic visual literacy.

ber of halftone dots per linear inch (lpi)—of the finest halftone in your collection. The higher the screen frequency, the finer the detail and the smoother the appearance of the image. Newspapers typically use an 85–100-lpi screen. Most halftones in monographs and serials are 120, 133, or 150 lpi, although some can be higher. Very high-quality promotional materials or fine art reproductions use 180 to 200 lpi or more. With the possible exception of the latter category, 600-dpi, 8-bit capture provides more than adequate resolution.[39] In reality, 400-dpi, 8-bit capture is probably sufficient, especially since the file size is less than half that of the 600-dpi file. Scanning a range of 19th- and early 20th-century halftones at 400 dpi, 8-bit, Cornell did not discern any noticeable moiré patterns.[40]

▸ Continuous tone or continuous-tone-like material—items such as photographs, watercolors, and some finely inscribed line art that exhibit smoothly or subtly varying tones. Tone and color reproduction are as important as detail or more so. Always scan this material at multiple bits. Experts suggest bit-depths greater than 8 per channel to capture the tonal information contained in photographs or to reproduce 8-bit/channel files that map more closely to human visual perception.[41]

▸ Mixed material—mixed documents, such as illustrated pages from a newspaper or book, that comprise more than one of the other categories. Mixed documents can be problematic to scan, as different regions of the page may call for different approaches. Duke University found this out in its sheet music project: they decided to image at 150 dpi, 24-bit color to retain legible text, which resulted in moiré in the cover graphics. A few companies have attempted to

39. Ironically, color halftones may be scanned at lower resolutions because the various color screens are placed at different angles, minimizing the visual impact of moiré patterns from one screen to the next.

40. The computer age introduced a range of halftones, many of which utilize random or irregular patterns and are therefore less challenging to image than conventional halftones.

41. Franziska Frey, "Digital Imaging for Photographic Collections: Foundations for Technical Standards," *RLG DigiNews* 1, no. 3 (December 15, 1997), www.rlg.org/preserv/diginews/diginews3.html; Sabine Süsstrunk, "Imaging Production Systems at Corbis Corporation," *RLG DigiNews* 2, no. 4 (August 15, 1998), www.rlg.org/preserv/diginews/diginews2-4.html.

build scanning systems that treat each region of a mixed page optimally. Other approaches call for grayscale capture at sufficient resolution, with postprocessing of various zones to optimize their presentation onscreen or in print.[42]

Medium and Support

Reflective documents may be distinguished further by the media and support used to create them. Both individually and collectively, these affect the digitization process. For instance, a printed text may not be a good candidate for bitonal scanning if it is produced on thinner paper with ink show-through or bleed-through, or if acid content has darkened the pages, reducing the contrast between medium and support.

The most common material for support is paper, but some institutions have imaged items produced on cloth, papyrus, palm leaf, parchment, vellum, and other animal skin. Paper can be absorbent, like newsprint paper, or reflective, like coated stock or photographic paper.[43] It can also have a matte or glossy finish, or be obviously textured or cockled. Decide whether to capture or minimize the support's physical properties and appreciate that scanners may amplify surface texture. Imaging papyrus, for instance, Finnish researchers found that the text was partially obscured by the background texture of the papyrus. On the other hand, scholars have noted that the Vatican Library Project images captured the texture of the parchment, enabling them to tell which side was written on.[44]

The medium used to record information also affects the digitization process. Some media, such as graphite pencil, early inks, and the gold leaf used to illuminate manuscripts, are highly reflective and more difficult to capture than more absorbing media, such as common printing inks or watercolors. Some pigments fall outside the gamut of current color spaces. Documents containing a range of media (e.g., pencil annotations over typewriting) pose other challenges, especially when one media tends to obscure the other. Consider too the density of the media. Is it light or dark? Does it vary across the page? What constitutes the lightest white? On typescript pages, correction fluid is brighter than the paper support. Some incised texts, such as palm-leaf manuscripts or stylus tablets, contain no media and may require specialized lighting or image processing to ensure good contrast between the strokes and the support.

Tonal Representation

Source documents may be characterized by their density or dynamic range—the range of tonal difference between the lightest light and darkest dark. Originals can theoretically exhibit a range of density values from 0 (no variation) to 4, indicating that the lightest portion is 10,000 times brighter than the darkest portion; in other words dynamic range is measured logarithmically. A densitometer measures dynamic range to determine the amount of light that is blocked in transmissive originals or absorbed in reflective originals. The higher the dynamic range, the more potential shades can be represented, although there is no automatic correlation between the dynamic range and the number of tones represented. For instance, high-contrast microfilm exhibits a broad dynamic range, but reproduces few tones.

Table 3.1 illustrates the typical limits of various document types. The density of reflective originals rarely exceeds 2.0, although the exact number varies with the material type or the gloss of the paper or inks used. Coated stock is generally less dense than photographic paper, but is denser than newsprint. Transmissive materials can exhibit a greater dynamic range than reflective originals, generally around 3.0, but again the numbers vary with film type, exposure, polarity, and processing. Although positive slide film can convey higher densities, several photographic and imaging experts argue for using negative film because it typically has a smoother response curve over its dynamic range, exhibiting more shades and greater shadow and highlight detail. It is also easier to work with.[45]

---

42. Picture Elements developed a halftone utility to detect and process halftone content. See Kenney and Sharpe, "Illustrated Book Study."

43. Elizabeth Lunning and Roy Perkinson, *The Print Council of America's Paper Sample Book: A Practical Guide to the Description of Paper* (Cambridge, MA: Print Council of America, 1996).

44. Antti Nurminen, "Recording, Processing and Archiving Carbonized Papyri," www.cs.hut.fi/papyrus/TOC.html; Henry M. Gladney, Fred Mintzer, and Fabio Schiattarella, "Safeguarding Digital Library Contents and Users. Digital Images of Treasured Antiquities," *D-Lib Magazine* (July/August 1997), www.dlib.org/dlib/july97/vatican/07gladney.html.

45. Ester, *Digital Image Collections*, p. 10; "KODAK PHOTO CD Technology/Transferring Images," *Kodak Service & Support*,

Table 3.1. Typical Dynamic Ranges for Source Documents[46]

Document	Dynamic Range
Newsprint	0.9
Printed material	1.5
Coated stock	1.5–1.9
Photographic prints	1.4–2.0
Negative films	2.8
Commercial grade colored slides	2.8–3.0
High grade transparencies	3.0–4.0

More projects have begun to measure the dynamic range of source documents to quantify their tonal characteristics. To date, these efforts have been confined to rare or special collections, but they may become more common. Several reports on digitizing photographs, illustrations, papyrus, and incunabula provide good information.[47]

If your institution does not have special equipment, table 3.1 can help you determine tonal resolution requirements. Also characterize your document's tonal composition subjectively and pinpoint the most challenging information to capture. If you are digitizing photographic prints, do the shadows or highlights contain important detail? Was the image properly exposed or will you be attempting to recover detail that is hidden or washed out? Scanning professionals refer to an image's key type. An image with detail concentrated in the highlights is called a high-key image;

one with extensive detail in the shadows is known as low-key; one with evenly distributed detail content is considered balanced (see figure 3.5). The most challenging to capture are low-key images, images that are underexposed, and those produced on transparent media. In general, black-and-white originals are easier to characterize than color ones, as hue and saturation tend to confuse the perception of brightness.

Dynamic range also describes a digital system's ability to reproduce tonal information. Theoretically, the broader a system's dynamic range, the smoother the transition between dark and light and the more tones it can represent. A system's ability to represent tones is most important for continuous-tone documents and for photographs it may be the single most important aspect of image quality.

Conventional wisdom is to select a scanning system that can match or exceed the dynamic range of the most demanding source documents. For instance, scanning systems with a lower dynamic range may capture reflective materials effectively but not transmissive ones.[48] Theoretically, then, the dynamic range of your source documents determines the minimum dynamic range of your scanner. Table 3.2 presents typical dynamic ranges for various scanning systems.

A Note of Caution   Correlating dynamic range with tone reproduction has several problems, however:

▸ Obtaining reliable information on a scanner's dynamic range can be difficult. Product literature often gives only a theoretical range, computed mathematically based on the bit-depth under ideal conditions rather than actual system performance.

▸ Bit-depth *loosely* correlates with tonal resolution but can be misleading about image quality. The upper bits of many scanning devices carry a declining percentage of reliable information

Frequently Asked Questions, www.kodak.com/global/en/service/faqs/faq1001c.shtml; Brian Lawler, *Optimizing Photo CD Scans for Prepress and Publishing* (Eastman Kodak Company, 1995), ftp://ftp.Kodak.com/pub/photo-cd/prepress/dci346.pdf.

46. Chart compiled from a number of sources: Robert G. Gann, *Desktop Scanners: Image Quality Evaluation* (Upper Saddle River, NJ: Prentice-Hall, 1999), p. 216; Blatner and Roth, *Real World Scanning,* pp. 26–27; Sybil Ihrig and Emil Ihrig, *Scanning the Professional Way* (Berkeley, CA: Osborne McGraw-Hill, 1995), p. 18; Philip Greenspun, "Adding Images to Your Site," *Philip and Alex's Guide to Web Publishing,* photo.net/wtr/thebook/images.html; Lawler, *Optimizing Photo CD Scans.*

47. Donald P. D'Amato and Rex C. Klopfenstein, "Requirements and Options for the Digitization of the Illustration Collections of the National Museum of Natural History" (March 1996), www.nmnh.si.edu/cris/techrpts/imagopts; Colet, Keller, and Landsberg, "Digitizing Photographic Collections"; Nurminen, "Carbonized Papyri"; F. C. Mintzer et al., "Toward On-line, Worldwide Access to Vatican Library Materials," *IBM Journal of Research and Development* 40, no. 2 (1996), www.research.ibm.com/journal/rd/mintz/mintzer.html.

48. This has led some to assert that reflective media, including photographic prints, can be captured easily by most midrange flatbed scanners with a dynamic range of 3.0. See Ihrig and Ihrig, *Scanning the Professional Way,* pp. 18–19; TASI provides a chart that correlates optical density, transmittance, contrast range, and approximate bit-depth equivalent: "Scanners," *TASI,* www.tasi.ac.uk/building/scannersprint.html.

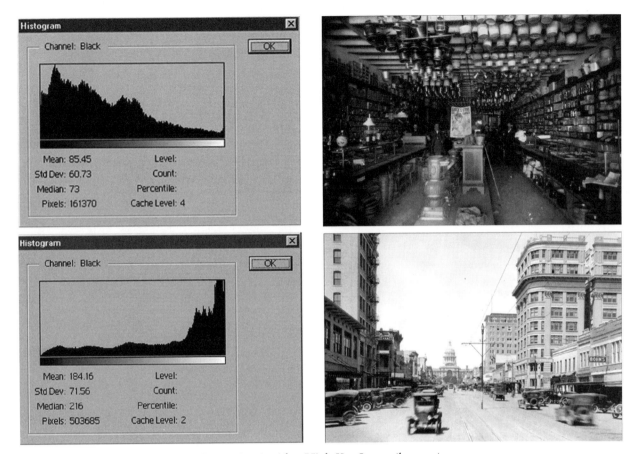

Figure 3.5 Comparing a Low-Key Image (top) with a High-Key Image (bottom)

Table 3.2. Typical Dynamic Ranges for Various Scanning Systems[49]

Device	Dynamic Range
Grayscale scanners	2.2 or below
Color flatbed scanners	2.5–3.9[50]
Drum scanners	3.4–4.0
Digital cameras	Based on ISO ratings[51]
Film/transparency scanners[52]	2.2–3.7

(that is, signal, as opposed to noise). Increased bit-depth does not automatically correlate to greater tonal resolution. A flatbed scanner with a lower bit-depth but a better signal-to-noise ratio may actually reproduce tones more accurately than one with greater bit-depth but a higher degree of noise. Therefore, to evaluate systems with the same stated bit-depth and dynamic range, ask about the signal-to-noise ratio.[53]

▸ Bit-depth is linear, but human perception is not. Increasingly, scanning systems capture at greater bit-depths internally (say 10–14 bits/channel) and then reduce to 8-bit data output that more closely matches to human perception.[54]

Good tonal reproduction, then, represents a

49. Compiled from information in Ihrig and Ihrig, *Scanning the Professional Way*, and product literature.

50. Older or low-end models have dynamic ranges around 2.5 or lower; midrange, about 2.8-3.2; high-end, in the 3.4-3.9 range.

51. As pointed out in Ihrig and Ihrig, *Scanning the Professional Way*, p. 33, digital cameras typically state dynamic range in terms of equivalent ISO ratings.

52. Low-end film scanners offer 2.2-2.8; midrange, 2.8-3.2; and high-end up to 3.7.

53. Ihrig and Ihrig, *Scanning the Professional Way*, p. 28. Gann, *Desktop Scanners*, chap. 15, describes various tests related to tonal resolution of a scanner.

54. Corbis and the LC Pictorial Imaging Project use this process. Several imaging experts claim that 12-bit linear data can be reduced to 8-bit nonlinear data with no perceptual loss if the reduction look-up table (LUT) adapts to the dynamic range of the data: Reilly and Frey, "Evaluation of Digital Images"; Donald S. Brown, "Image Capture Beyond 24-Bit RGB," *RLG DigiNews* 3, no. 5 (October 15, 1999), www.rlg.org/preserv/diginews/diginews3-5. html.

Interpreting Digital Scanner/Camera Specifications

Don Williams
Image Scientist, Image Engineering and Simulation Lab, Image Science Division, Eastman Kodak Company

So, you are in the market for a digital scanner or camera. You have a rough idea of your image quality and workflow needs. You may have even spoken to peers or read reviews about existing devices. While correct in thinking that generally "you get what you pay for," you want to avoid excess and constrain costs. Hands-on experience would be the best selection method, but since manufacturers are not in the habit of providing loaners for evaluation, you turn to their specification literature to help you decide. But what do these specifications mean?

Resolution, File Size

Manufacturers of scanners for document imaging (e.g., flatbed scanners, copy-stand digital cameras, or microfilm scanners) specify resolution in terms of the number of dots, samples, or pixels per linear inch (dpi, spi, or ppi). Two flavors are often cited: optical (or addressable) resolution and interpolated resolution. Optical resolution indicates the actual sampling rate at the instant that the optical imaging sensor scans the document. Some manufacturers using linear sensor arrays cite optical resolution in each of the two directions of a scan (i.e., 400 × 600 dpi). The lower number is associated with the sensor's pixel pitch; the higher one with the scanner's stepping action in the other direction. The lower number is often a truer indicator of performance since the sensor itself usually contributes more imaging performance.

For equivalent dpi, prefer optical resolution to interpolated resolution. Interpolated resolution is usually the result of scanning at a lower optical resolution and mathematically filling in pixels between native pixels. Practical interpolation methods do not perfectly predict these filled-in pixels and can only be considered guesses of the real document information. Manufacturers often inflate their products' performance by citing interpolated resolution values, which carried to an extreme (usually anything above 1,000 dpi on a document) can be misleading.

Keep in mind that resolution specifications merely indicate how frequently the document is sampled (i.e., the spatial sam-

pling rate). Extraction of the smallest details depends on how light energy spreads in the capture process, which is measured by line spread function (LSF) or the Modulation Transfer Function (MTF).[1]

Because there is no established document object reference for digital cameras, product literature refers to resolution in a number of different ways. For instance, a camera with 2,048 line × 3,072 pixel resolution is often referred to as a 6-Megapixel (Mpixel) camera (2,048 × 3,072 = 6,291,456). Sometimes the uncompressed, finished file size, which includes the total bit-depth for all channels, is given as another measure of camera resolution (2,048 × 3,072 × 24 bits)/8 bits per byte = 18 MB). For lower-resolution cameras, under 1 Mpixel, specifications are sometimes given in terms of equivalent monitor resolutions; for instance, VGA (640 pixels × 480 lines) or SVGA (800 pixels × 600 lines) resolution.

Digital cameras that use full-frame color filter area arrays (CFA) employ interpolated spatial resolution. This type of interpolation is much better than that used in most document scanners, which attempt to fill in pixels. Digital cameras interpolate missing color pixels with correlated knowledge of the color actually sampled at that location. Since resolution for a given pixel correlates well between color channels, CFA sensors can achieve very good, almost lossless interpolation.

Bit-Depth

Determine first whether the product vendor is referring to finished-file bits or internal bits. For instance, some color document scanners boast 30-bit color depth (10 bits/color channel × 3 colors), yet only yield a 24-bit color finished file. In this case, 10 bits/channel are acquired at initial capture, but are reported as only 8 bits/channel after the internal image processing. Though claiming the higher of the two has always been common engineering practice, it has recently become a marketing practice, sometimes without caveat. Bitonal or 1-bit/channel scanners, for example, have always captured at 8

CONTINUED ON PAGE 41

combination of scanning bit-depth, signal-to-noise ratio, and final bit-depth output.

Mapping Tones Correctly   The first consideration in translating tonal information from analog to digital is to ensure that the scanning device can represent the full dynamic range of the source document. The second consideration is to ensure that tonal values have mapped correctly. Scanning hardware and software use a histogram to analyze the tonal character of the original in preview mode. Histograms also help determine how the

values are distributed throughout an image and whether any have been stretched or clipped in the highlight, shadow, or midtone regions. You can use scanner controls or imaging software to adjust the histogram to emphasize an image's most important features. For instance, if the shadows contain key information, you can shift the tonal curve to redistribute tonal values to the darker portion of the image, while compressing the levels in the lighter portions (see figure 4.3, page 78). Gamma values (or brightness settings) are a way to represent the overall tonal distribu-

CONTINUED FROM PAGE **40**

bits/channel and used this bit-depth to make intelligent image processing decisions to yield high-quality 1-bit finished files. However, some binary scanner specifications are now citing 8-bit depth (internal), which at first reading may be mistaken for grayscale output.

Dynamic Range, Maximum Density

Dynamic range is the useful range of densities that a capture device can detect. Of all the camera and scanner specification claims, it is perhaps the most misstated. Nearly all manufacturers calculate dynamic range values under the absolute best capture conditions.[2] These conditions rarely, if ever, exist in practice. For 8-, 10-, 11-, and 12-bit per channel devices, the calculated dynamic ranges are 2.4, 3.0, 3.3, and 3.6 respectively. So be cautious whenever these particular dynamic ranges or maximum densities are cited. They are probably the result of a theoretical calculation, not based on true system performance. In fact, dynamic range specifications somewhat removed from these numbers, say 3.2 or 2.8, are probably more believable.

Occasionally, manufacturers state different dynamic range and maximum achievable density values. Think of dynamic range as a viewing window into a world of all possible densities. The window's width is defined by the dynamic range. In theory, you can see all possible densities by moving the window. For instance, a manufacturer might cite a dynamic range of 3.3 accompanied by a maximum density of 3.4. The difference of 0.1 is merely a density bias indicating that density levels between 0.0 and 0.1 will be given the same digital count value. The manufacturer has set the minimum window boundary at 0.1 rather than 0.0. Since the window width remains the same, the maximum density moves from 3.3 to 3.4.

Three Examples

1) Sometimes it is fun to gather details in a specification and crosscheck them for compatibility: A manufacturer of a 4,096-lines-x-4,096-pixels, three-color camera claims a 48-MB file from a 12-bits/color channel capture. But doing the math shows a 72-MB file instead.

$$File\ size = \frac{(4,096 \times 4,096\ pixels) \times 12\ bits/channel \times 3\ colors}{8\ bits/byte} = 72\ MB$$

What gives? Substituting 8 bits/channel instead of 12 results in the 48-MB file indicated in the specification. This suggests that the 12 bits/color is internal, and only 8 bits/color are in the finished 48-MB file.

2) The resolution section of a digital camera specification claims, "Reproduction quality superior to 4 × 5 film—no film grain."

Cute wording! Since no film is being used for detection, there will, of course, be no film grain. This doesn't mean that there will not be other grain or noise sources due to the electronic detector.

3) From a film scanner specification:

Optical density	3.6 dynamic range
A/D conversion	12 bits
Output data	8 bits or 16 bits per color channel
Multisample scanning	Scan 4 or 16 times for reduced noise
Scan time	Approximately 20 seconds

Dynamic range likely assumes no losses due to optical flare. The 12-bit data can be scaled to an 8-bits/color finished file or stored as a 16-bit file. The extra 4 bits/pixel of the latter are probably empty imaging information.

The multisample scanning actually works very well for reducing scanner-related noise. It will not reduce film-grain noise though and you are likely to increase the scan time.

Notes

1. Don Williams, "What is an MTF . . . and Why Should You Care?" *RLG DigiNews* 2, no. 1 (February 15, 1998), www.rlg.org/preserv/diginews/diginews21.html.

2. A theoretical dynamic range value can be calculated with respect to the bit-depth per channel, N, and reported in density units as

$$Dynamic\ Range = -log10 \frac{(1)}{(2^N-1)}$$

The statements herein are the opinions of the author and not Eastman Kodak Company.

tion numerically, as represented in the histogram.[55]

Typically, scanning technicians set aimpoint values (numeric representations) for white point that correspond to the lightest portion of the image or grayscale target, and gamma values to reflect the overall composition of the source document. The Library of Congress and the National Archives have defined scanning aimpoints for white and black at levels slightly inside the maximum values. In the digital world, black is defined as 0 (or 100% black) and white is 255 (or 0% black). NARA chose values of 8 (or 97% black) for black and 247 (3% black) for white—leaving some headroom to ensure no loss of detail or clipping in scanning. A grayscale target enables you to anchor aimpoint values for black, white, and neutral gray to a source with known digital values and to use that information to set scanner controls, calibrate scanners and monitors, and conduct quality

55. Corbis applies gamma corrections during scanning to compensate for under- or over-exposed images or to capture high- or low-key scene content. Süsstrunk, "Imaging Production Systems." See also Don Williams's sidebar on image-quality metrics.

control. NARA and LC also assigned tolerance levels—or acceptable deviations from those set values. NARA's tolerance level for grayscale scanning was more strict (at ± 1% variation) than for color scanning (±3 RGB levels), because variance is more noticeable and more easily controlled in grayscale. Finally, NARA set a maximum tonal range, recognizing that many documents, especially low-contrast ones, would have considerably fewer tones than the maximum.[56]

These values are helpful in establishing measurable and verifiable requirements, but vendors and institutions alike still struggle with notions of acceptable quality. John Stokes of JJT Inc., who won the contract to scan pictorial material for the Library of Congress, observed that even though the aimpoint values were met and three different JJT staff members conducted subjective quality review, LC reject rates remained high until they initiated periodic joint image review.[57] Imaging by the numbers is particularly problematic with negatives. Sometimes, precise requirements can't make an image look right. Dependence on technical definitions may sacrifice quality to consistency. As James Reilly has noted, the goal is not to reproduce the grayscale bar, but to convey the essence of the original object or to render intent—what the photographer meant to convey rather than what is recorded on the photographic medium.[58]

Tonal reproduction is still difficult to define precisely but it is receiving a great deal of attention. A growing number of institutions argue strongly for the use of grayscale targets with every scan and the recording of density values in the digital files to facilitate the use of tonal metric information in subsequent presentation and preservation of digital materials. At least one institution also recommends the recording of objective information about the lighting properties under which the document was scanned, to enable a printer or monitor to reflect similar tonal values.[59]

Color Appearance

Color evaluation is the most challenging part of document assessment because color is difficult to judge and varies with the viewing conditions. Even scientifically, color properties are difficult to quantify, and it is easy to be dissatisfied with color rendering without being able to describe objectively what's wrong. Color's hue (red, green, blue, etc.), saturation (intensity), and brightness (lightness) depend on the level of illumination and type of light. Characterizing color in these three dimensions is the first step. Then you must translate between analog and digital, between one color space and another, between reflected and transmitted light and back again. All these steps are bound to introduce inaccuracies in color appearance.

Begin assessing color by clarifying color's role in representing the informational content of the originals. Answer these questions to determine conversion requirements:

1. *Is color reproduction necessary to the document's meaning?* Some essentially monochrome documents may have had color introduced through age, use, or improper processing. In the LC manuscript digitization project, subject specialists distinguished between intended and unintended color. Something like yellowing paper did not warrant color capture. But if primary or secondary creators introduced color (e.g., a production company's red underlining on a typed script), it was considered significant.

2. *What is the nature of the color?* Is the color flat and the palette limited, as in a poster, or continuous and highly varied, as in a photograph? The more variation and range represented by the color, the more difficult it is to convey.

3. *What purpose does the color serve?* In some cases, such as artwork or photography, the colors of the original are significant in themselves, for either aesthetic or informational reasons. Color also encodes information; for instance, maps frequently use color to differentiate topographic or physiographic features. Sometimes,

56. Puglia and Roginski, "NARA Guidelines"; Library of Congress, "RFP97-9 Table of Contents." A good description of setting white point, gamma, and black point is provided in "Workflow Guidelines: 2. Pre-scanning Preparation and Post-scanning Processing of the Originals," *TASI*, www.tasi.ac.uk/building/workflow2.html and www.tasi.ac.uk/building/workflow3.html.

57. John Stokes, "Imaging Pictorial Collections at the Library of Congress," *RLG DigiNews* 3, no. 2 (April 15, 1999), www.rlg.org/preserv/diginews/diginews3-2.html.

58. Digital Imaging for Libraries and Archives Workshop (Digitizing Photographs session), Cornell University Library, Ithaca, NY, July 1998.

59. Psohlavec, "Digitization of Manuscripts," p. 23.

as in pie charts, the meaning of the color could be represented another way. Must you capture colors?

4. *How important is maintaining the color appearance?* Define your tolerance for color matches. While it may be critical to capture the actual shades of blues favored by Chagall, you might be satisfied with a digital representation that approximates the blue of veins in a medical illustration.

Measuring Color   Well-defined standards for measuring color and appearance include both instrumental and visual techniques. The American Society for Testing and Materials maintains several hundred standards of their own, and other organizations, such as ISO and the Technical Association of the Pulp and Paper Industry (TAPPI), are also major contributors to this field.[60] The human eye should always be the final arbiter, even when color instruments are used. However, both visual assessment and color vocabulary can be highly subjective—what is meant by dark red, for instance? Further, unless you are particularly experienced, it is difficult to translate between color perception and the numeric values assigned to the three dimensions of color. Several tests enable visual specification of color difference. One using the Munsell® Book of Color compares original materials to a set of color test samples. The color samples most closely representing the color properties of the original are assigned numeric values developed in the Munsell color-order system.[61] The key to visual color assessment is to anchor the object to some control sample (a color bar or grayscale target) that has known color numeric values.

Specialized instruments (spectrophotometers or colorimeters) measure the color properties of an original objectively. These instruments standardize color assessment with numeric values and, along with software, translate between color space and the various color systems. GretagMacbeth has developed a color evaluation method that is used extensively by a range of organizations and manufacturers and that could be adapted for imaging programs. The method combines instrumental and visual analysis, including recommendations on equipment, viewing conditions, color standards, tolerances, and color analysis.[62] The Mitretek® Systems' report to the Smithsonian's National Museum of Natural History presents an excellent description of characterizing cultural documents with color instruments.[63]

Color assessment is problematic, but capturing and conveying color appearance is arguably the most difficult aspect of analog or digital imaging. In the digital realm, color reproduction depends on variables such as the level of illumination at the time of capture, the capabilities of the scanning system, and mathematical representation of color information across the digitization chain. Applications such as printing have addressed faithful color appearance fairly successfully, however almost any report on digitization of cultural resources confirms the difficulty and frustration of capturing and presenting color that approximates the original. Among problems encountered are:

▸ Some older materials are produced with pigments that are not accurately represented in color models designed for digital imaging, which are optimized for the colors of photographic dyes. As a result, a few projects have developed their own imaging systems. To help the Vatican image centuries-old manuscripts, IBM developed a scanner with a customized colorimetric filter set. The Visual Arts System for Archiving and Retrieval of Images (VASARI) project and its successor, the Methodology for Art Reproduction in Color (MARC) project, developed specialized digital imaging systems to capture exceptionally accurate color for conservation, scientific analysis, and art reproduction. The process requires extremely tight control on all phases of the imaging chain and results in very large image files (up to 800 MB).[64]

▸ Some institutions have also established custom color profiles for scanning to avoid adjusting each image individually. MOMA, for example,

60. American Society for Testing and Materials, *ASTM Standards on Color and Appearance Measurement*, 5th ed. (West Conshohocken, PA: ASTM, 1996).

61. Ibid., "Standard Practice for Specifying Color by the Munsell System, D1535-95B," pp. 52–79.

62. "Instrumental Quality Control," sect. 5 in *Fundamentals of Color and Appearance* (New Windsor, NY: MacBeth, 1996).

63. D'Amato and Klopfenstein, "Digitization of the Illustration Collections."

64. "VASARI: Visual Arts System for Archiving and Retrieval of Images," www.ecs.soton.ac.uk/~km/projs/vasari.html.

created color profiles for the various photographic media in their collection.[65]

▸ Color has both color (chrominance) and specular (luminance) aspects, so it is important to define color under particular viewing conditions. By controlling the lighting at the time of capture, you can use the spectral information to correct color error in the scan and duplicate the lighting conditions for image viewing; otherwise, you can spend significant time correcting color during quality control.[66] Vatican Library Project staff used information about lighting at capture to correct the image onscreen and convey a spatially uniform illumination. Institutions often use illumination that approximates noon daylight appearance, although MOMA captured material under conditions that match exhibition lighting.[67]

Although no hard-and-fast rules control color appearance, experts say:[68]

▸ Characterize the color requirements based on document assessment. If using a transparency, consider any color casts attributable to the film.

▸ Establish a tolerance range for color shifts or variation from aimpoint values.

▸ Calibrate the scanning device to a color standard, using standard color and grayscale targets and software. Software can characterize deviations from known target values and create device-specific profiles to adjust the color of scanned images accordingly.[69]

▸ Calibrate the rest of the imaging system to the scanner. Consider use of color management system (CMS) software to achieve device-independent color consistency.

▸ Control the lighting and environment for image production and evaluation.

▸ Scan grayscale and color targets with each image or production batch.

▸ Evaluate the color appearance of the displayed image against the source document.

▸ Stay close to the scanning device's capabilities and make only minor corrections; avoid postprocessing of the master image.[70]

▸ Consider capturing files in greater than 8-bits-per-channel color or producing 8-bit/color output from a richer bit level.

▸ Save the image in RGB or CIE color space, if possible with International Color Consortium (ICC) profiles.

▸ Save the image as a TIFF file with color profile information.

▸ Maintain scanning-related metadata.

Detail

Detail seems to be one of the most easily measurable properties and for some categories of materials, this is true. Detail capture in scanning is governed by resolution, bit-depth, and system performance.

Measuring Detail

▸ *Printed text*: Measure the height of the smallest lowercase letter that typifies the item or group of items. Use an eye loupe (10X) with a scale reticle with increments to the nearest .1 mm.

▸ *Manuscripts, line art, and other hard-edged representations*: Measure the finest stroke-width that must be represented and characterize the needed level of quality (e.g., barely detected, fully reproduced, etc.). Use a powerful eye

65. Colet, Keller, and Landsberg, "Digitizing Photographic Collections," pp. 5, 11.

66. This was discovered by the Johnson Art Museum at Cornell University in their digital imaging project. Issues associated with color fidelity are discussed in Gann, *Desktop Scanners*, pp. 249–264.

67. If using a flatbed scanner, determine the spectrum of the light source; the best is D_{65} or daylight 6500 kelvin lighting. Colet, Keller, and Landsberg, "Digitizing Photographic Collections," p. 10.

68. Michael Ester describes standard practice at Luna in Ester, *Digital Image Collections*, pp. 14–15. Ihrig and Ihrig, *Scanning the Professional Way*, pp. 55–76, offers excellent advice on calibration, controlling the environment, and use of color management systems. The California Heritage Digital Image Access Project used targets, data gathering, and analysis techniques developed in-house that mimic what a color management tool could do: "Digitizing the Collection," *California Heritage Collection*. "Image Capture and Manipulation: Guidelines and Procedures," *TASI*, www.tasi.ac.uk/building/image_cap1.html.

69. Two tests for color fidelity are recommended in Gann, *Desktop Scanners*, pp. 260–263 and appendix A.

70. Frey and Reilly, *Digital Imaging for Photographic Collections*, p. 40; D'Amato and Klopfenstein, "Digitization of the Illustration Collections." Project staff at the Blake Archive do individual color corrections against the original transparencies, which can take anywhere from 30 minutes to several hours: *The William Blake Archive*, www.iath.virginia.edu/blake.

loupe or microscope (minimum 30X) with a light source and scale reticle to the nearest .01 mm.

▸ *Photographs and other continuous-tone documents*: Measure the finest scale or line that should be represented in the digital image, such as a telephone line, an individual hair, or fringe on a hat. Measure a sharp or crisp detail rather than a fuzzy or poorly defined one. If you use film formats, place them on a light box and measure the finest scale using a microscope with a magnification of at least 50X.

▸ *Halftones*: Measure the screen frequency with a screen finder. Place it over the halftone at an angle and move it until a four-pointed moiré star points to the screen count it has measured. A publication can contain a range of halftone screen sizes, so take several measurements.

▸ *Mixed items*: Take measurements on various portions of a mixed page. For instance, if an item contains text and halftones, measure the smallest character and the halftone screen ruling.

Defining Detail as Character Height The ANSI/ AIIM preservation microfilming standard for determining requirements for text legibility is based on a quality index (QI) method.[71] The standard assumes that detail is a fixed metric—the height of the smallest significant lowercase letter in a document determines the necessary resolution required for a predictable level of quality.

The QI method forecasts the levels of image quality (e.g., tolerances), from barely legible through excellent rendering. Excellent rendering of a lowercase character requires 8 line pairs; good legibility could be ensured at 5 line pairs; 3.6 line pairs provide marginal quality; and 3.0 means that the character is barely legible. The QI method uses technical test targets to relate text legibility to system resolution. In microfilming, this is expressed as QI = p × h. "QI" equals levels of quality; "p" equates resolution to the number associated with

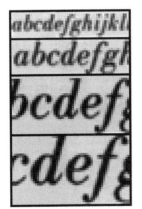

5 pixels over height of e

8 pixels over height of e

12 pixels over height of e

16 pixels over height of e

Figure 3.6 Character Representation Based on Number of Pixels

the smallest line pair pattern/mm distinguishable on a test target; and "h" equals the height of the smallest character (measured in mm).

The C10 Standards Committee of AIIM developed a digital QI formula, which Cornell University further refined for scanning brittle books. The first formula was designed for bitonal scanning, and provides for a generous sampling requirement to compensate for misregistration and reduced quality due to creating only black and white pixels:

$$QI = (dpi \times .039h)/3$$
$$h = 3QI/.039dpi$$
$$dpi = 3QI/.039h$$

Because tonal images subtly "gray out" pixels that are only partially on a stroke, a separate formula applies to grayscale scanning of printed text:[72]

$$QI = (dpi \times .039h)/2$$
$$h = 2QI/.039dpi$$
$$dpi = 2QI/.039h$$

In the early 1990s, Cornell discovered that a 1-mm-high character was a good anchor for resolution requirements for digitizing brittle books published from 1850–1950. Virtually no publishers during this period used fonts shorter than 1 mm. Although the pages of the brittle books might have darkened slightly, in general the text was clean and bitonal scanning captured it effectively. The bitonal

71. Association for Information and Image Management, *Practice for Operational Procedures/Inspection and Quality Control of First-Generation, Silver Microfilm of Documents* (Silver Spring, MD: Association for Information and Image Management, 1998); Nancy Elkington, ed., *RLG Preservation Microfilming Handbook* (Mountain View, CA: Research Libraries Group, 1992).

72. Anne R. Kenney and Stephen Chapman, *Tutorial. Digital Resolution Requirements for Replacing Text-Based Material: Methods for Benchmarking Image Quality* (Washington, DC: Commission on Preservation and Access, 1995).

Figure 3.7 Adequately Rendered Stroke (see also plate 2)

Figure 3.8 Inadequately Rendered Stroke

QI formula predicted that the textual information could be captured with excellent quality at a resolution of 600 dpi. At this resolution, a 1-mm-high character could be rendered bitonally with excellent fidelity if 24 pixels spanned its height; if captured in grayscale, a 1-mm-high character would be rendered with excellent fidelity if 16 pixels spanned its height. This figure is analogous to the 8 line pairs required in microfilming (each pair equals 2 lines, for a total of 16 lines).

An extensive onscreen and print examination of digital facsimiles for the smallest Roman and non-Roman fonts used during this period confirmed these benchmarks. Although many of the books do not contain such small text, to avoid an item-by-item review, all books were scanned at 600 dpi.

Columbia University also benchmarked against 1-mm-high text in their oversize map project, but captured the documents in 24-bit to convey their color characteristics. Columbia determined that

Table 3.3.   Quality Index for Stroke Rendering

Pixels	Quality Assessment
≥2	Excellent/high fidelity
1.5–1.9	Good
1–1.5	Questionable; should be assessed onscreen
< 1	Poor to unacceptable

adequate legibility could be assured at 200 dpi.[73] At that resolution, a 1-mm-high character is covered by 7.8 pixels. According to the quality index approach for multibit imaging, this would provide legible but not faithful rendering (e.g., a QI of 4). Both the National Archives and the Library of Congress have recommended scanning maps at 300 dpi for access; this resolution equals about 12 dots across a 1-mm-high character (e.g., a QI of 6).

Defining Detail as Stroke   The QI method was designed principally for machine-produced textual documents. For handwritten correspondence or nontextual documents containing distinct edge-based graphical representations, such as maps, sketches, etchings and engravings, a fixed reference such as character height cannot so easily characterize detail. For many such documents, a better representation of detail would be the width of the finest line, stroke, dot, or marking that must be captured in the digital surrogate.

Engineers and imaging scientists seem to agree that in digital imaging the finest feature should be covered by at least two pixels to be fully repre-

73. Janet Gertz, *Oversize Color Images Project, 1994-1995, A Report to the Commission on Preservation and Access* (Washington, DC: Commission on Preservation and Access, 1995), p. 8, www.columbia.edu/dlc/nysmb/reports/phase1.html.

sented. This would mean, for instance, that an original with a stroke measuring $1/100$ inch (.254 mm) must be scanned at a rate of at least 200 dpi to resolve its finest feature. For bitonal scanning, this requirement would be higher (say 3 pixels/feature), due to the potential for sampling errors and the conversion to black and white pixels. A feature or stroke can often be *detected* at lower resolutions, on the order of 1 pixel/feature, but at this point issues of quality tolerance come into play. While investigating the use of the KODAK Photo CD™ technology to image special collections materials, Cornell confirmed that 2-pixel coverage provided adequate stroke rendering (see figure 3.7).[74] Cornell also attempted to define quality projections at lower resolutions. In figure 3.8, for example, the borders of the document were inadequately rendered, with at most a single pixel straddling the finest line. Although not definitive, table 3.3 equates pixels to perceived levels of quality in stroke rendering.

Defining Detail as Scale   Imaging scientists analyze scale space by defining the dominant scale, the distance covered by the most significant, often repeated, structural feature; for example, knots in a carpet. Different illustration processes exhibit different dominant scales; for example, an etching can be distinguished from an engraving by the characterization of the process used to create it. Imaging scientists have explored a range of techniques for analyzing scale space. If the different scales in a document can be measured, it is possible to develop benchmarks for digital resolution requirements. However, Cornell and Picture Elements rejected this approach in the Illustrated Book Study when it was revealed that the finest scale for some illustrations would require imaging at resolutions above 1,000 dpi.[75]

Assessing detail in photographs and other continuous-tone documents can be equally problematic. James Reilly suggests that detail may be defined as "relatively small-scale parts of a subject," but this requires an assessment of the finest scale.[76] At the granular level, photographic media are characterized by random clusters of irregular size and shape, with no distinct boundaries, which can be practically meaningless or difficult to distinguish from background noise. Further, capturing that level of granularity requires an extremely high resolution. And meaningful level of detail in a photograph is both subjective and variable. Perhaps cityscape street signs visible under magnification should all be clearly rendered in the digital surrogate, but what about individual hairs or pores in a portrait? Many institutions have based their resolution requirements for photographic media on the quality that can be obtained from prints generated at a certain size (e.g., 8" × 10") from a certain film format (e.g., 35 mm, 4" × 5").

Defining Detail Based on Visual Perception   It is possible to peg resolution requirements to visual perception, even though a document may contain detail invisible to the unaided eye. A person with 20/20 vision can discern detail that is approximately $1/215$ inch wide under normal lighting conditions at a standard reading distance. The human eye optically averages details finer than this. If it takes at least two pixels to fully render the finest observable stroke, then a scanning resolution at twice visual perception (e.g., 215 × 2 = 430) in grayscale or color should preserve details at levels detected by the human eye. Cornell and Picture Elements adopted this approach in the Illustrated Book Study. Although some book illustrations contained exceedingly fine detail, the project's scholarly advisory committee found that 400-dpi, 8-bit capture distilled the hidden structural features surprisingly well. The images replicated the detail present in the original even when viewed close up or under slight magnification (e.g., 2X). Robert Thibadeau and Evan Benoit, working with antique books, observed this same visual phenomenon although they recommended 600-dpi capture for letterpress books to render print imperfections satisfactorily, if not completely accurately.[77] Other projects have concluded that increasing resolution improves quality appreciably up to a certain point—between 400 and 600 dpi (grayscale or color)—increasing resolution beyond that must be weighed against rising costs or larger file sizes.

---

74. Kenney and Rieger, *Using Kodak Photo CD Technology*.
75. Kenney and Sharpe, "Illustrated Book Study."
76. Reilly and Frey, "Evaluation of Digital Images," p. 20.

77. Kenney and Sharpe, "Illustrated Book Study"; Thibadeau and Benoit, "Antique Books."

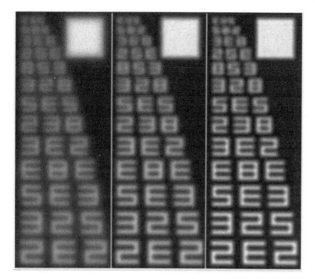

Figure 3.9 RIT Alphanumeric Test Object Scanned on Three Different Systems

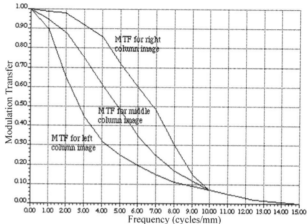

Figure 3.10 MTF Values for Figure 3.9

Translating between Digital Resolution Requirements and Scanner Performance   Detail capture is governed by resolution, bit-depth, and system performance. Typically, cultural institutions have stated resolution requirements in terms of dots per inch or pixels per inch (ppi). But dpi does not by itself sufficiently measure system performance. A higher-dpi scanner may actually deliver a poorer image than a high-quality scanner with a lower dpi. Dots per inch indicates a scanner's sampling interval (e.g., 300 dpi) but this does not guarantee accurate capture of details as fine as $^{1}/_{300}$ inch.

"Limiting resolution" is another way to state resolution, commonly used with bitonal scanning. This measurement is based on the finest feature that a system can detect in a scanned image of a test target. Limiting resolution assumes that if the finest feature is detected, then the quality of the rest of the image will follow. This may not be the case—especially for grayscale or color scanning—as three versions of the RIT target illustrate (see figure 3.9). The finest feature has been detected in each case, but the image quality differs significantly.

For grayscale or color images, Modulation Transfer Function (MTF) analysis is a superior way to determine scanning system performance. MTF analysis correlates known details (frequencies) with how well the system detects those details. The MTF is not a single number, but a continuum of measurements over increasingly finer details. Don

Williams of Eastman Kodak has provided an excellent introduction to MTF in an article for *RLG DigiNews*.[78]

MTF has been used extensively by the image evaluation community, but MTF targets and the software to calculate the MTF generally have been available only recently.[79] A number of institutions, including the Library of Congress, are beginning to consider MTF requirements in developing requests for proposals for imaging services.[80] Some government agencies have adopted it in such specialized domains as fingerprint analysis, military reconnaissance, and medical imaging.[81]

MTF indicates how well a scanning system is working, but we need a way to translate between a determination of pixel coverage or dpi to an MTF requirement. In other words, we need the means for defining quality requirements based on our documents that ultimately results in system performance measurements. The process is problematic, in part because the cultural community has not established imaging standards and because MTF requirements can be difficult to determine over all spatial frequencies for a range of document

78. Don Williams, "What Is an MTF . . . and Why Should You Care?" *RLG DigiNews* 2, no. 1 (February 15, 1998), www.rlg.org/preserv/diginews/diginews21.html.

79. Gann, "Recommended Test" and "Applied Image Test Target," in *Desktop Scanners*, pp. 183–199, 280–282.

80. Library of Congress, "RFP97-9 Table of Contents"; Stokes, "Imaging Pictorial Collections."

81. Federal Bureau of Investigation, Criminal Justice Information Service, *Electronic Fingerprint Transmission Specification (EFTS)* (Washington, DC: CJIS, 1999), appendices F and G. MITRE provides free MTF software: "Image Quality Evaluation," www.mitre.org/technology/mtf.

types. But we can begin by understanding how the limiting frequency can be set.

Consider for instance the following: If two pixels are necessary to cover the finest stroke on a document and that stroke measures .1 mm, two pixels cover .1 mm or 20 pixels cover 1 mm, which translates to 512 dpi, converted from metric to English measurement. For the sake of this argument, round down to 500 dpi. To ensure that the scanner can capture information correctly at that level, the resolution requirements need to combine 500 dpi *plus* a high MTF performance. Assume that the MTF analysis uses a sine-wave test target and that a frequency in the test target corresponds to the finest feature of the document. Begin by equating digital resolution to classical resolution by dividing the dpi by 2 and converting from inches to millimeters. At 500 dpi, 250 line pairs per inch (lppi) convert to a metric measurement of 9.84 lppm (250/25.4 = 9.84 lppm) or approximately 10 cycles per millimeter (cpm).

If 10 cpm represents the finest feature to detect, what is the MTF value? The ISO/FDIS 12233 indicates that the limiting resolution is the spatial frequency corresponding to a camera output modulation level of 5%, which would represent an MTF value of .05.[82] In its RFP for Pictorial Imaging, LC suggested MTF values of .2 to 1 for the finest frequency sampled, which was 10 cpm. This same value was chosen by MITRE to define the lowest MTF value for fingerprint analysis for the FBI. Don Williams suggests that when he and his colleagues assess scanners, they look for MTF values between .1 and .4 at the finest frequency they want to detect. Figures higher than that might indicate the presence of sampling artifacts. So, should the lowest MTF value be set at .05, .1, or .2? The values closer to .1 will provide a smoother appearance, which might be appropriate for continuous-tone materials. Where hard-edged detail is especially important, as for the FBI fingerprint requirements, choose a higher value, say .2.[83] In the final analysis, a subjective criterion determines that value.

Once you have determined the limiting resolution and the required level of capture, the next step is to define values for other frequencies up to a point where the MTF value would be 1.0 (indicating that the detail has been fully preserved in the image).[84] In general, the higher the value, the better the preservation of detail. LC drew on the MITRE specifications and an assessment of the performance of scanners tested at Image Permanence Institute and Rochester Institute of Technology.[85] The general idea is to define values that conform to the informational content of the source document and also reflect human visual perception. For instance, an MTF for the image can be overlaid with one that has been defined based on human perception to arrive at a subjective image-quality requirement, or it can be used in tandem with known aspects of display technology.[86] The chart in figure 3.10 plots the values into MTF curves for each scanner. The curve for the right-hand image remains higher longer than the curves for the two other scanners. This is clearly a good curve for this image, which is based on hard-edge detection. It may not be a good curve for continuous-tone information, however. Curve characterizations are needed for a range of document types, as well as the appropriate targets and software to characterize a range of scanning systems based on those interpretations.

Is MTF ready for prime time? Although MTF has been applied successfully elsewhere, it seems cultural institutions and vendors alike are still uncertain just how to use it. In 1997, the Library of Congress became the first major cultural institution to require MTF targets and analysis in its RFP bids and vendor evaluation. The chosen vendor could not meet the MTF specifications at the required resolution of 5,000 pixels and ended up capturing at 7,500 pixels and then scaling down to 5,000. For its part, the Library of Congress did not consider the MTF values a requirement of the contract for production

82. "Resolution Test Chart," in "Program of Work," *PIMA/IT10: Electronic Still Picture Imaging*, www.pima.net/standards/it10/IT10_POW.htm#12233.

83. Even when values are set, it is important to consider whether a true signal is being detected, since the signal-to-noise ratio will affect both detection and estimation of the signal.

84. Values above 1.0 indicate that the image has been sharpened; above 1.5–1.8, that it has been over-sharpened to the point that noise produces false reads (Don Williams, telephone conversation with Anne R. Kenney, April 22, 1999).

85. James Reilly, e-mail to Anne R. Kenney, April 19, 1999; Franziska Frey, e-mail to Anne R. Kenney, April 21, 1999.

86. There are a number of weighted subjective image-quality metrics, such as acutance, subjective quality factor (SWF), and square root integral (SQRI): Don Williams, "An Overview of Image Quality Metrics," *RLG DigiNews* (forthcoming) www.rlg.org/preserv/diginews.

activity and did not use the MTF values as part of its inspection process.[87] Nonetheless this was an important first step and work is underway to facilitate the use of MTF analysis in the very near term.[88]

Determining Conversion Requirements for Printed Output

Editors, prepress houses, video and slide production services, and, increasingly, cultural repositories base their conversion requirements on their need for printed output. Their resolution requirements depend less on informational content than on the printing devices. To establish resolution requirements using this approach, determine the physical dimensions of the original, media and dimensions of the largest output, bit-depth for scanning, whether the output will be treated as text or line art, continuous-tone or halftone, and the resolution and/or screen frequency of the printing device. Although formulas can lead to precise resolution requirements, industry experts say to scan at integral resolutions of the scanner's optical resolution (and scale down later if necessary) and base enlargement factors on an integral of the original document's dimensions.[89]

Text or Line Art Output to a Printer

To capture text or line art bitonally, the general rule of thumb is to match the resolution of your printer. For enlargements, increase the scanning resolution. Ihrig and Ihrig, among others, offer this formula: Scanning resolution = printer resolution × enlargement factor. They also suggest that resolutions (including interpolated resolutions) up to 1,200 dpi can smooth out jagged lines. Jeff Bone argues that resolutions above 600 dpi may not be discernible and the difference in imaging and printing time will be considerable.[90]

Grayscale/Color Documents Output to a Continuous-Tone Printer or Recorder

Specialized printers, such as dye-sublimation printers or film recorders, print tonal information dot-for-dot rather than producing halftone screens. These are most often used to print continuous-tone images, especially photographs. If you use a continuous tone-printer, the general recommendation is to print dot for dot, as determined by the printer's resolution. Also consider any level of enlargement. The same formula used for text and line art applies to continuous tone printing, although the resolution of these printers is usually much lower. A number of institutions base their imaging requirements on a printed output of a certain size; for instance, 300 dpi for an 8" × 10" print, which would require a 300-dpi scanning resolution for an 8" × 10" original, but a 600-dpi scan of a 4" × 5" original. You could use a pixel resolution instead, stated as 3,000 pixels along the long dimension of each format.

A few industry experts suggest that lower scan resolutions can produce excellent results, especially if the original images exhibit soft features or the printer's output is meant to be viewed from a distance. Jeff Bone has developed a rather complex color scanning formula for continuous-tone output that takes into consideration the recorder's resolution, the dimensions of the original, the enlargement factor, and a quality coefficient (QC). The quality coefficient ranges from normal (1) for soft-featured photographs with slight pixelation that may be evident up close, to ultimate (3.3) or pixel-for-pixel scans designed for 35-mm and 4" × 5" transparencies, which he suggests would only be used for testing.[91]

Grayscale/Color Documents Output to a Halftone Printer

Most printers impart a halftone screen to grayscale or color image files and the prepress industry has developed fairly consistent formulas to determine input scanning requirements for a predetermined output. These formulas take into consideration the size of the original, the enlargement factor, the resolution and screen frequency of the printer, and the level of gray or color values to be represented. Scanning resolutions for halftone reproduction are

87. Stokes, "Imaging Pictorial Collections"; Phil Michel, e-mail to Anne R. Kenney, March 19, 1999. The FBI discovered that a number of scanners used for fingerprint scanning had to sample at a much higher rate and then subsample to meet the MTF requirements as defined in FBI, *Electronic Fingerprint Transmission*, appendix G. Don D'Amato, telephone conversation with Anne R. Kenney, July 1, 1999.

88. "Program of Work," *PIMA/IT10*.

89. Ihrig and Ihrig, *Scanning the Professional Way*, p. 94.

90. Ibid., p. 98; Jeff Bone, "Line Art Scanning," *The Scanning FAQ*, www.infomedia.net/scan/TSF-LineArt.html.

91. Jeff Bone, "Color Scanning," *The Scanning FAQ*, www.infomedia.net/scan/TSF-Color.html.

usually pegged at 1.5 or 2 times the resolution of the screen frequency of the printer (lpi). Halftone printers that use modern techniques including frequency-modulated screening, which relies on random or irregular patterns of variably shaped dots, do not require this "fudge factor." For printing devices using conventional screen technologies, with a regular pattern and fixed dot sizes, a common formula is dpi = lpi × enlargement factor × 1.5. You can also use this formula to determine your printer requirements; for instance, if you have scanned a photograph at 300 dpi, you can produce a 100-lpi halftone print up to twice the size of the original. Jeff Bone's scanning FAQ Web site includes a Scanning Resolution Calculator that you can download.[92]

You can determine the spatial resolution that can be represented, but what about the tonal resolution? Scanning publications provide a formula for determining the maximum number of distinct tonal levels that halftones can represent, based on the printer's resolution and the halftone screen frequency: Maximum number of tonal values = (printer resolution ÷ lpi)^2 + 1. For instance, using this formula you can calculate that you need a 2,400-dpi-enabled printer to reproduce 256 tones per color at line screens up to 150.[93]

Other Considerations: Enhancement, File Formats, and Compression

Enhancement/Image Processing

There is some debate within the cultural community about enhancements in the master file. Purists insist on storing images as raw capture data and treating only access files. They argue that enhancements are irreversible and image quality may suffer as the files are migrated. The change may be negligible at first, but it is cumulative and will ulti-

mately severely compromise the master file. Others justify minor enhancements in the master file to improve the quality or decrease costs under strictly controlled and well-documented circumstances.

Generally accepted enhancements or image-processing routines to the master file include:

▸ Reduction of greater than 8-bit/channel linear data to 8-bit nonlinear data

▸ Contrast stretching

▸ Minimal adjustments for color and tone

▸ Descreening/rescreening of halftones

Debatable image processing techniques include:

▸ Image sharpening

▸ Despeckling for bitonal images

▸ Deskewing

▸ Software-controlled color/tone enhancement

▸ Application of color management profiles

▸ Conversion to CMYK or sRGB color

File Formats

The file format for master images should be able to preserve the resolution, bit-depth, color support, and metadata of a very rich image file. It should also handle being stored either uncompressed or compressed using both lossless and lossy techniques. The format should be open and well-documented, widely supported, and cross-platform compatible. Table 3.4 summarizes important attributes for the seven most common image formats in use today. Despite a good deal of interest in finding alternative formats for masters, TIFF meets most of these requirements and remains the de facto standard. An October 1998 survey including responses from 25 Department of Defense and other federal records officers revealed that all use TIFF, with 73% also reporting the use of PDF files.[94] As of this writing, however, the National Archives has no immediate plans to develop guidelines for federal agencies creating image files and

92. Ihrig and Ihrig, *Scanning the Professional Way*, pp. 103–105, discusses the rule of 2:1 versus 1.5:1 and suggests using the lower number. Agfa suggests using 1.5 for screen rulings greater than 133 lpi and 2 for screen ruling of 133 or less; for stochastic halftone printing, scanning at the conventional screen ruling: Agfa, *An Introduction to Digital Scanning* (Belgium: Agfa-Gevaert, 1994), p. 21. Jeff Bone, "Greyscale Scanning," *The Scanning FAQ*, www.infomedia.net/scan/TSF-GreyScale.html; Jeff Bone, "Scanning Resolution Calculator," *The Scanning FAQ*, www.infomedia.net/scan/calcs.html.

93. Blatner and Roth, *Real World Scanning*, p. 18; "Information: Color Management," *Dunaway Products*, www.dunaway.com/Secties/Information/ColorManagement/ColorManagement.html.

94. Sue McTavish, "DoD-NARA Scanned Images Standards Conference," *RLG DigiNews* 3, no. 2 (April 15, 1999), www.rlg.org/preserv/diginews/diginews3-2.html.

Table 3.4.   Common Image File Formats[95]

Name and Current Version	TIFF 6.0 (Tagged Image File Format)	GIF 89a (Graphics Interchange Format)	JPEG (Joint Photo-graphic Expert Group) /JFIF (JPEG File Inter-change Format)	Flashpix 1.0.2
Extension(s)	.tif, .tiff	.gif	.jpeg, jpg, .jif, .jfif	.fpx
Bit-depth(s)	1-bit bitonal; 4- or 8-bit grayscale or palette color; up to 64-bit color[96]	1-8 bit bitonal, gray-scale, or color	8-bit grayscale; 24-bit color	8-bit grayscale; 24 bit color
Compression	Uncompressed Lossless; ITU G4, LZW, etc. Lossy: JPEG	Lossless: LZW[97]	Lossy: JPEG Lossless:[98]	Uncompressed Lossy: JPEG
Standard/Proprietary	De facto standard	De facto standard	JPEG: ISO 10918-1/2 JFIF: de facto standard[100]	Publicly available specification
Color Mgmt.[103]	RGB, Palette, YC_bC_r,[104] CMYK, CIE L*a*b*	Palette	YC_bC_r	PhotoYCC and NIF RGB,[105] ICC (optional)
Web Support	Plug-in or external application	Native since Micro-soft® Internet Explorer 3, Netscape Navigator® 2	Native since Micro-soft® Internet Explorer 2, Netscape Navigator® 2	Plug-in
Metadata Support	Basic set of labeled tags	Free-text comment field	Free-text comment field	Extensive set of labeled tags
Comments	Supports multiple images/file[107]	May be replaced by PNG; interlacing and transparency support by most Web browsers	Progressive JPEG widely supported by by Web browsers[108]	Provides multiple resolutions of each image; wide industry support, but limited current applications
Home Page	home.earthlink.net/ ~ritter/tiff (unofficial)	Specification: cica.cica. indiana.edu/graphics/ image_specs/gif.89 .format.txt	www.jpeg.org/public/ jpeghomepage.htm	www.digitalimaging.org

Table 3.4   (continued)

ImagePac, Photo CD	PNG 1.2 (Portable Network Graphics)	PDF 1.3 (Portable Document Format)
.pcd	.png	.pdf
24-bit color	1-48-bit; 8-bit color, 16-bit grayscale, 48-bit color color	4-bit grayscale; 8-bit color; up to 64-bit color support
Lossy: "Visually lossless" Kodak proprietary format[99]	Lossless: Deflate, an LZ77 derivative	Uncompressed Lossless: ITU G4, LZW Lossy: JPEG
Proprietary	ISO 15948 (anticipated)[101]	De facto standard[102]
PhotoYCC	Palette, sRGB, ICC	RGB, YC_bC_r, CMYK
Java™ applet or external application[106]	Native since Microsoft® Internet Explorer 4, Netscape® Navigator 4.04, (incomplete in late 1999)	Plug-in or external application
Through external databases; no inherent metadata	Basic set of labeled tags plus user-defined tags.	Basic set of labeled tags
Provides 5 or 6 different resolutions of each image; unclear future	May replace GIF	Preferred for printing and viewing multipage documents; strong government use
www.kodak.com:80/ US/en/digital/ products/photoCD .shtml	www.cdrom.com/ pub/png	www.adobe.com/ prodindex/acrobat/ adobepdf.html

Notes to Table 3.4

95. Table compiled by Richard Entlich, Department of Preservation, Cornell University Library; will be maintained as part of Cornell University, Department of Preservation, "Cornell Online Imaging Tutorial," www.library.cornell.edu/ preservation.

96. Though the TIFF 6.0 specification provides for 64-bit color, many TIFF readers support a maximum of 24-bit color.

97. LZW is patented and its use in software development may require licensing and royalty payments: "LZW Patent and Software Information," *Unisys*, www.unisys.com/unisys/lzw.

98. A non-JPEG compressed file in a JFIF wrapper. A specification for a lossless JPEG (JPEG-LS) has not been finalized. "JPEG—Information Links," *JPEG and JBIG Webpages*, www.jpeg.org/public/jpeglinks. htm.

99. Visually lossless compression techniques are themselves lossy, but take advantage of characteristics of human sight to create an image that is virtually indistinguishable from its uncompressed form.

100. C-Cube Microsystems released JFIF into the public domain. The "official" file format for JPEG files is SPIFF (Still Picture Interchange File Format), but by the time it was released, JFIF had already achieved wide acceptance. SPIFF, which has the ISO designation 10918-3, offers more versatile compression, color management, and metadata capacity than JPEG/JFIF, but it has little support. It may be superseded by JPEG 2000/DIG 2000: "JPEG 2000 and the DIG: The Picture of Compatibility," www.digitalimaging.org/i_dig2000. html.

101. Approved by W3C to replace GIF for Web use.

102. Theoretically, Adobe has released enough information to allow developers to write applications that read and modify PDF files. In practice, PDF files are usually created and accessed using Adobe's own Acrobat® software.

103. See "Color 101," page 64, for definitions of terms used in this category.

104. Like CIE Lab, YC_bC_r is composed of three channels: one for luminance (Y) and two for chrominance (CC).

105. The Flashpix 1.0.2 specification defines NIF RGB and sRGB identically. sRGB is still in development. The next revision of the Flashpix specification may move to sRGB if it is stabilized by then.

106. Kodak makes available a set of server-side CGI scripts that convert PCD images to JPEGs for native viewing in Web browsers: "Technical Description," *Kodak: PhotoCD on the Web*, www.kodak.com:80/global/en/ professional/support/PCDWeb/techDescription.shtml. However, viewing in this manner sacrifices access to the ImagePac's special features.

107. The TIFF 6.0 specification calls for the ability to store multiple TIFF images in a single file, but not all TIFF readers support this feature.

108. Some versions of Microsoft® Internet Explorer may not display progressive JPEGs properly.

will not accept digital images into its Center for Electronic Records.

The Arts and Humanities Data Service has developed guidelines for depositors that include technical standards governing preferred and acceptable formats. Preferred formats correspond to archive formats for long-term storage; acceptable formats may need to be reformatted to the preferred format during accessioning. Of the four AHDS service providers that accept image files for deposit, three list only TIFF in the preferred format category, and one (History Data Service) also accepts PNG and SPIFF files for long-term retention.[109] The Technical Advisory Service for Images (TASI) favors PNG and SPIFF because they are open formats, offer good metadata capability, and use lossless compression. Unfortunately, support for these formats has been slow in coming. For the time being, TASI acknowledges that TIFF remains a safer choice, but does recommend migrating to these formats as use increases.[110] Interest in the Flashpix file format, which has wide industry support but limited current applications, is growing. Corbis will adapt its production workflow to take advantage of the format, although it will continue to scan all original photographs in TIFF format for the time being.[111] Finally, a Basic Image Interchange Format (BIIF) is in development that may well offer the most robust potential for long-term management.[112]

Compression

Just as it debates enhancement, the cultural community continues to discuss the use of compression in the master file. Some argue for storing the master file uncompressed to facilitate digital preservation and for limiting compression to access versions. Others support the use of a standard, lossless compression to save space, arguing that its effects are reversible. Still others suggest that balancing quality with cost justifies a modest level of lossy compression.[113] Critics of this approach, however, argue that whenever you touch a lossy file you may lose information—a circumstance that could increase as files are migrated to new formats and new forms of compression.

Lossless compression, particularly ITU Group IV and sometimes JBIG, is routinely used to achieve significant savings in file size in bitonal image master files (ranging from 10:1 to 30:1). JSTOR and other initiatives have also turned to Cartesian Perceptual Compression (CPC), achieving even greater storage savings (on the order of 100:1). Cornell and Xerox Corporation are codeveloping DigiPaper, an image-based document representation that uses token-based image processing to obtain very high compression without sacrificing image quality.[114] For tonal images, lossless compression schemes often do not result in significant savings (e.g., 2.5:1), hence the interest in employing lossy schemes—from those that are "visually" lossless (Photo CD) to controlled levels of JPEG compression at levels of 5:1 or 10:1. Proponents of this approach differentiate between textual and continuous-tone files and between grayscale and color images. They also carefully specify the processes used in applying compression to maintain color representation, and minimize data loss.[115] Whatever today's choice, however, it seems certain that new file formats and compression processes will completely change the information landscape in the future.

Conversion Guidelines from Various Institutions

No "right" choices govern image conversion because guidelines depend on institutional settings, priorities, and resources, and may not scale beyond those constraints or to other environments. Benchmarking allows you to step through a process that leads to the creation of your own spe-

109. Neil Beagrie and Daniel Greenstein, "Managing Digital Collections: AHDS Policies, Standards and Practices Version 1" (December 15, 1998), *Arts and Humanities Data Service*, ahds.ac.uk/public/srg.html.

110. "Selecting Formats," *TASI*, www.tasi.ac.uk/building/choosingprint.html.

111. "Corbis and Live Picture to Bring Flashpix™ Format to Stock Photography" (April 23, 1997), *MGI Software Corp./Live Picture Inc.*, livepicture.com/press/releases/p28.html.

112. ISO/IEC, "IS 12087-5, Information Technology—Computer Graphics and Image Processing—Image Processing Interchange (IPI)—Functional Specification—Part 5: Basic Image Interchange Format (BIIF)" (December 1, 1998), 164.214.2.51/ntb/baseline/ docs/biif/index.html.

113. Robin Dale, "Lossy or Lossless? File Compression Strategies Discussion at ALA," *RLG DigiNews* 3, no. 2 (February 15, 1999), www.rlg.org/preserv/diginews/diginews3-1.html.

114. "Cartesian Perceptual Compression: Technical Information," *Cartesian Products, Inc.*, www.cartesianinc.com/Tech; *DigiPaper*, www3.cs.cornell.edu/DigiPaper. DigiPaper also facilitates functions such as image-to-text conversion.

115. D'Amato and Klopfenstein, "Digitization of the Illustration Collections"; Mintzer et al., "Vatican Library Materials."

cific guidelines. In this context, other institutions' conversion guidelines are worth reviewing. The Cornell University Library Department of Preservation maintains on its Web site a chart of conversion guidelines from representative institutions.[116]

BENCHMARKING FOR ACCESS

Using the Web to make retrospective resources accessible to a broader public raises issues of image quality, utility, and delivery at the users' end. Libraries and archives must evaluate the effects of technical choices made in creating *and* presenting digital images.

Why benchmark for digital access if the materials are converted properly? The primary reason is that if digital masters are created for the long haul, they are not very user friendly given current technical and cost constraints:

▸ Derivatives are used in various ways (browsing, detailed review, printing), so high resolution in one context may be poor quality in another.

▸ Access dictates other requirements, such as speed of delivery, technical capabilities, user tolerance, and costs. In other words, a good access image conveys the most desired information given the constraining influence (e.g., file size, cost, etc.).

▸ Users' needs may conflict with what is affordable and what the technology can deliver. Unlike digital masters, access images may be short-term products.

User studies have concluded that all researchers expect fast retrieval, acceptable quality, and added functionality from digital images. Of course, they want lots of other things, too, such as the ability to print, manipulate, annotate, compare, and contrast images and, increasingly, specialized services. This leads cultural institutions to confront a whole host of technical issues that don't exist in the analog world. Unfortunately, the cumulative effect of technological choices on the transmission and display of digital material has not been systematically assessed. File formats, compression processes, scripting routines, transfer protocols, Web brow-

sers, processing capabilities, and the like will combine to affect user satisfaction, especially given the lag in users' adoption of technology. Users may want the highest quality, but may be frustrated by how long it takes to download a file, or disappointed when a beautiful color image displays in a largely posterized form.[117]

Speed of Delivery

Users probably care most about speed of delivery. Several variables control access speed, including the file size, network speed and traffic, and the time to read the file from storage and to open it on the desktop.[118] Table 3.5 provides common data transfer rates based on a range of network bandwidths.[119]

Table 3.5. 1999 Network Operating Speeds (in MB/second)

OC-48 (Abilene backbone)	300
vBNS (NSF/MCI backbone)	77.8
FDDI	12.5
100BaseT Ethernet	12.5
DS-3 (T-3)	5.6
10BaseT Ethernet	1.25
Cable modems	up to .47 downstream
ADSL	.19–1 downstream
DS-1(T-1)	.192
ISDN (home use)	.018
v.90 modem	.007

You can calculate a maximum transmission rate in seconds (t): t = number of bytes per file ÷ [bandwidth (in bytes/sec.) × .8]. A 1-MB file might be accessed in a tenth of a second by FDDI (fiber),

---

116. Cornell University, Department of Preservation, "Cornell Online Imaging Tutorial," www.library.cornell.edu/preservation.

117. Posterization is the appearance of visible tonal steps. For example, if a 24-bit image is viewed on an 8-bit monitor, reduction of a 16.7 million-color palette to a 256-color one causes posterization. Also, redistribution of grayscale values during tonal correction may cause posterization by leaving unused gray levels.

118. John P. Weise, "Data Transfer Time," *Preparing Quality Images for Computer Networks*, www-personal.umich.edu/~jweise/quality/DataTransferTime.HTML; Cornell University, "Online Imaging Tutorial."

119. Chart prepared from "Frequently Asked Questions," *The Abilene Project*, www.ucaid.edu/abilene/html/faq-general.html; "Frequently Asked Questions," *vBNS*, www.vbns.net/press/press_faqs.html; "Networking Terms," *University of Georgia, University Computing Networking Services*, www.uga.edu/~ucns/lans/glossary.html. In January 1999, a portion of the NSF/MCI backbone also started carrying traffic at OC-48 speed, which is four times faster than the rest of the backbone.

but takes nearly three minutes on a v.90 modem. Because they cannot control network connections, cultural institutions have focused on constraining image file size to speed access and some institutions have determined file size limitations for network access. The Etext Center at the University of Virginia, for example, keeps lower-quality images under 100 KB and the higher-quality images at 300–500 KB. The Thomas A. Edison Papers project at Rutgers University aims to deliver the content of the images "legibly at a minimum file size." Staff found that on busy days file sizes over 60–80 KB could drag even over fast connections. Timely delivery has dictated access only to items that can be delivered legibly at that file size.[120]

Typically institutions have reduced file size by reducing the resolution or bit-depth or by applying compression. Each of these choices can have a pronounced impact on image quality.

Resolution

Most master files will be scaled to produce lower-resolution display versions. Reducing the resolution reduces file size geometrically (e.g., halving the resolution reduces file size by 75%) and also enables more of the image to appear onscreen. Institutions typically reduce resolution either to the pixel dimensions of common desktop display settings or to a standard dpi.

Bit-Depth

Decreasing the bit-depth of multibit files reduces file size; ironically, bitonal master files are usually scaled for access by *increasing* the bit depth. Bit-depth reduction can be done with an image-editing program or by converting to a new file format.[121] The National Archives chose bit-depth reduction to minimize the end user's computing and monitor requirements. Tests from workstations not on the New York Public Library intranet indicated that load times were too long, so the NYPL chose to deliver several access versions of stereoscopic views in grayscale rather than color. In general, however, bit-depth reduction—in particular the use of GIF—has been used more to address variation in system palettes among monitors or insufficient video memory to support full color images.

Compression

Many institutions compress access images to speed delivery, opting for JPEG compression for multibit images to control the level of compression. In the past, it took a long time to decompress images at the desktop, but this is becoming less of a concern as RAM increases. The main concern in using high compression levels is image quality, and some institutions have sacrificed speed to minimize annoying compression artifacts. JPEG artifacts are especially evident in text, but a poorly compressed JPEG image of a photograph can look worse onscreen than a properly created lossless GIF. Highly efficient compression schemes that enable the delivery of large images over slow networks while retaining image quality and offering multiple views to the end user are gaining popularity.[122]

Multi-Image Files

Although bundled images may not increase the initial speed of delivery, they can facilitate flipping through a cache of downloaded images. The most notable example of this is PDF for multipage documents, although some institutions have experimented with the use of multiple-image TIFF files, MPEG, QuickTime® movies, and CPC compression.

Increasing Bandwidth

The need to reduce file size to speed delivery may decrease as broad-bandwidth information pipelines and wireless high-speed data transfer capabilities are developed in the next five to ten years to support research, electronic commerce, and entertainment. Current FCC rules require all analog broadcasts to be phased out by the end of 2006. The potential of digital television, in particular

120. Bob Rosenberg, e-mail to Anne R. Kenney, April 5, 1999.

121. The University of Virginia discovered that there was no file-size gain in saving images compressed with JPEG at 8 bits instead of 24.

122. A questionnaire intended to help choose the appropriate compression process is offered in Association for Information and Image Management, *Selecting an Appropriate Image Compression Method to Match User Requirements* (Silver Spring, MD: Association for Information and Image Management International, 1998). "Information Architecture White Paper IA-6801: Electronic Image Formats and Compression Algorithms," Rev. 2.0 (October 20, 1998) *Los Alamos National Laboratory, Information Architecture*,www.lanl.gov/projects/ia/stds/ia680120.html.

123. "Digital Television Consumer Information" (November 1998), *FCC, Office of Engineering and Technology*, www.fcc.gov/Bureaus/Engineering_Technology/Factsheets/dtv9811.html.

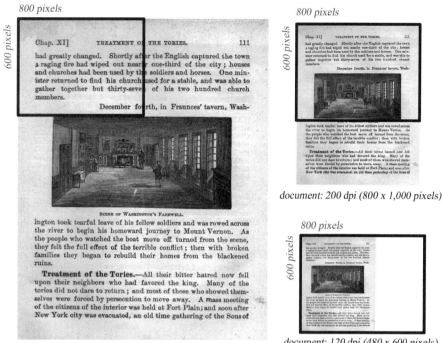

document: 4" x 5", 400 dpi (1,600 x 2,000 pixels)

document: 200 dpi (800 x 1,000 pixels)

document: 120 dpi (480 x 600 pixels)

Figure 3.11 Digital Image Resolution Compared with Display Resolution at 800 × 600 Pixels

Table 3.6. Percent of Image Displayed at Common Desktop Settings

Image Resolution for 4" × 5" Page	Desktop Settings		
	640 × 480 pixels	800 × 600 pixels	1,024 × 768 pixels
400 dpi (1,600 × 2,000)	10%	15%	25%
200 dpi (800 × 1,000)	38%	60%	77%
100 dpi (400 × 500)	96%	100%	100%
80 dpi (320 × 400)	100%	100%	100%

high-definition television (HDTV) is tantalizing.[123] The US government is funding efforts to build the Next Generation Internet (NGI) to link research labs and universities to high-speed networks that are 100 to 1,000 times faster than the current Internet. The NGI will make access to digital image files commonplace and high-quality audio and moving-image transfer practical.[124]

Completeness

Users expect to display the complete image at about the same size and with the same detail as the original. It is certainly possible to scale an image to

124. Elizabeth Cohen, "Internet2 and the Next Generations," *NARAS Journal* 8, no. 2 (winter 1998/99): pp. 51–56.

be completely represented onscreen at the same size as the original document, but this does not guarantee a legible image. In contrast to scanning devices and printers, current monitors offer relatively low resolution. Typical monitors support resolutions, or desktop settings, from a low of 640 × 480 to a high of 1,600 × 1,200, referring to the number of horizontal by vertical pixels painted on the screen when an image appears. Product literature often gives monitor resolutions in dpi, which can range from 60 to 120 or higher. In reality, any monitor that supports a range of desktop settings supports a corresponding number of screen resolutions (dpi). Switching to a higher setting (e.g., from 640 × 480 to 1,024 × 768) decreases the size and spacing between pixels and the monitor's dpi goes up. In other words, a "standard" 72-dpi display occurs only when the number of pixels displayed horizontally ("p") divided by the display window's width ("w") equals 72 (dpi = p/w).

So, how does a digital image's resolution relate to a display resolution? Consider the digital image of an illustrated text page and how best to display it. Let's say the original page was 4" × 5", scanned at 400 dpi in 8-bit grayscale. The pixel dimensions of this image are (4 × 400) by (5 × 400), or 1,600 × 2,000—above any desktop setting currently available. On a monitor set to 800 × 600, less than a sixth of the image will appear onscreen (see figure 3.11). A high-end monitor with a resolution of 1,600 × 1,200 would be able to display the full width of the image, but just over half of its length at once. This problem only increases for higher-resolution scans or images of larger-sized pages. For instance, under 2% of a 600-dpi image of an 8" × 10" document displays on a monitor set to 800 × 600. Table 3.6 illustrates how much of a sample illustrated text page displays at different resolutions at common desktop settings.

As table 3.6 indicates, if the sample page is scaled to 80 dpi, it displays completely at common desktop settings. This resolution ensures the complete display for similarly shaped pages whose height is less than or equal to the sample page. Documents greater than 6 inches in height or wider than 8 inches do not display completely. Determining the resolution of scaled images by dpi, then, does not guarantee that the complete image displays for all documents.

For the full image to display at a given desktop setting, the pixel dimensions of the image must be less than or equal to the pixel dimensions of the setting. Images can be scaled by setting one of their pixel dimensions to the corresponding pixel dimension of the desktop.[125] To fit the complete image from the sample page on a monitor set at 800 × 600, scale the vertical dimension of the image to 600; the corresponding horizontal dimension would be 480 to preserve the aspect ratio of the original. Reducing the 1,600 × 2,000-pixel image to 480 × 600 discards 91% of the informa-

tion in the original. This has the advantages of facilitating browsing with the full image and decreasing file size and transmission time. The downside should also be obvious: a major decrease in image quality as a significant number of pixels are discarded.

Image Quality

Onscreen image quality, including legibility and tonal reproduction, depends on variables associated with the image file, monitor and system capabilities, color representation, and the programs used to create derivative images.

Detail

Legibility and completeness often conflict. Cornell has developed benchmarking formulas for the display of text-based materials that correlate image quality, resolution, and the required level of detail:

$$dpi = QI/(.03h)$$
$$QI = dpi \times .03h$$
$$h = QI/(.03dpi)$$

In the formula, "dpi" refers to the resolution of an image (not to be confused with the monitor's dpi); "h" refers to the height of the smallest character (in mm); and "QI" refers to levels of legibility. This formula presumes that bitonal images are presented with at least three bits or more of gray and that filters and optimized scaling routines improve image presentation. Using this formula, you need to establish your own levels of acceptable quality. Cornell benchmarks readable text at a QI of 3.6, although 3.0 often suffices for cleanly produced text, particularly if it has been scanned in grayscale or color.

For example, for a 4" × 5" text page that contains a 1-mm-high character, set the QI requirement at 3.6. Use the benchmarking formula to predict the scaled image dpi:

$$dpi = QI/.03h, or$$
$$dpi = 3.6/(.03 \times 1), or$$
$$dpi = 120$$

The pixel dimensions for the scaled image would be (120 × 4) by (120 × 5), or 480 × 600. This full image could be viewed legibly on monitors set at 800 × 600 or above. If, on the other hand, you have determined that a QI of 3.0 offers

---

125. The formula for scaling for complete display of an image onscreen is: a. When digital image aspect ratio ≤ screen aspect ratio, set the image's horizontal pixel dimension to the screen's horizontal pixel dimension; b. When digital image aspect ratio is > screen aspect ratio, set the image's vertical pixel dimension to the screen's vertical pixel dimension.

sufficiently legible text, then the image could be delivered to the user at 100 dpi, representing pixel dimensions of 400 × 500.

The benchmarking formula can determine a preset scaling dpi based on document attributes or the needs of a particular clientele. In the Making of America project, Cornell scaled journal pages to 100 dpi to ensure legibility of the finest text. For all but the largest journals, this displays the full width of the page at the 800 × 600 setting, so users need to scroll in only one dimension.

But consider a scenario where your primary users rely on monitors with 640 × 480 desktop settings. They might be satisfied with viewing only half of the document if it meant that they could read the smallest type, which might occur only in footnotes. If your users are more interested in quick browsing, you might want to benchmark against the body of the text, rather than the smallest typed character. For instance, if the main text were in 12-point type and the smallest "e" measured 1.6 mm in height, then the sample page could be sent to the screen with a QI of 3.6 at 75 dpi (300 × 375 pixels)—well within the capabilities of the 640 × 480 setting.

Monitor and System Capabilities

In addition to resolution, other attributes will affect image quality, including bit-depth, monitor calibration, and size, image sharpness, refresh rate, and color spectrum of display. Insufficient video memory, for example, limits how much gray or color information a monitor can represent at any pixel dimension. Compensation by dithering the image can result in unwanted color shifts or posterization. To determine how much VRAM is needed, use this formula: VRAM requirement = (number of pixels displayed × bit-depth)/8. For example, to display a 24-bit image at SVGA resolution requires $(1{,}024 \times 768 \times 24)/8 = 2.36$ MB of VRAM.[126] Higher resolutions will require more memory. Increasingly, computers come bundled with 4 MB of VRAM or more, but check the specifications before buying. For many computers you

cannot increase the VRAM later without changing the entire video interface (e.g., adding a different video card).

A monitor may have sufficient resolution and bit-depth, but if improperly calibrated it affects the quality of access images. Several institutions, including the National Archives, the Denver Public Library, the University of Virginia, Luna, and Corbis, have gone so far as to create electronic targets and specify monitor settings to help users calibrate their monitors. Evidence suggests, however, that few users take advantage of these offerings. As Michael Ester notes, "The only controls that are apt to see widespread use are those that are built into applications and underlying software."[127]

Color Representation   Color appearance is most problematic, since it is affected by different browsers, underpowered monitors, and transfer between color spaces. Several Web sites provide useful information on Web palettes for access.[128] Some institutions are turning to file formats such as PNG, which supports both a Web-safe palette and sRGB, designed to ensure color consistency across platforms. Others are including grayscale/color targets with their images to enable the end user to adjust the color. Solutions may be forthcoming from industry as color representation is a growing concern in electronic commerce—no mail order business can afford to handle too many returns because the color of a shirt does not match its picture in the catalog, whether in print or on the Web.

Scaling Programs   The program and scripts used for the scaling will affect the quality of the presentation, particularly for illustrations such as halftones. Scaling can introduce moiré that requires special processing, as JJT Inc. discovered in preparing derivative images for the Library of Con-

---

126. Increasingly, monitors are representing "true color" at 32 bits, which may or may not affect the amount of VRAM needed to support 24-bit images: see Don Brown on monitor bit-depth, "FAQs," *RLG DigiNews* 3, no. 5 (October 15, 1999), www.rlg.org/preserv/diginews/diginews3-5.html.

127. Ester, *Digital Image Collections*, p. 16.

128. Many Web sites cover the topic of browser palettes: Both Lynda Weinman, "The Browser-Safe Color Palette," www.lynda.com/hex.html, and Victor S. Engel, "The Browser Safe Palette (BSP)," the-light.com/netcol.html, have links to other useful sites on browser palettes and other Web graphics–oriented material. See also John Weise, "DMS Guide to Web Color: Palettes & Gamma," *OIT Digital Media Solutions*, www.oit.itd.umich.edu/projects/DMS/answers/colorguide; "WWW Incorporation Case Study," *Digital Images in Multimedia Presentation (DIMP)*, www.ilrt.bris.ac.uk/mru/dimp/section4/web_case-prt.html.

gress.[129] Scaling programs are also used to reduce the bit-depth of an image and different processes result in substantially different quality. Pick a routine that works best based on the attributes of the images to be scaled so as to achieve the most compact file size while retaining the necessary tonal information. Some Web sites provide helpful information on scaling programs, optimizing graphics, and choosing file formats to enhance image quality.[130]

Balancing Speed, Completeness, and Image Quality

Clearly, for many categories of documents, it can be hard to satisfy readers' requirements for speed, completeness, and image quality, given the limitations of screen design, computing capabilities, network speeds, and Web browsers. Benchmarking

screen display must take all these variables into consideration along with the attributes of the digital images themselves.[131] The conflict between image size, legibility, and speed of delivery has led many institutions to create a variety of access images: thumbnails for browsing and onscreen comparison, full-screen images for legibility, and detailed views for intensive study. As the MESL final report revealed, institutions vary widely in their choice of image views, even among such standard ones as thumbnails.[132] Most agree on the desirability of multiple access images, but specific sizes and techniques will change with time. Cornell University Library's Department of Preservation maintains a chart of guidelines for access images from representative institutions.[133] And in the future, more access images will be created on demand from the master files.

129. See Stokes, "Imaging Pictorial Collections," and "Sheet Music Scanning Procedures," *Historic American Sheet Music, Duke University, Rare Book, Manuscript and Special Collections Library*, scriptorium.lib.duke.edu/sheetmusic/procedure.html, on processing to eliminate/reduce moiré in their sheet music project.

130. Patrick J. Lynch and Sarah Horton, "Web Style Guide," *Yale-New Haven Medical Center, Center for Advanced Instructional Media*, info.med.yale.edu/caim/manual/contents.html; "Graphics Viewers, Editors, Utilities and Info," *Hallym University Department of Physics*, physics.hallym.ac.kr/resource/hotlist/graphics.html; "Image File Formats List," *Center for Innovative Computer Applications*, cica.indiana.edu/graphics/image.formats.html; Mintzer et al., "Vatican Library Materials"; Kenney and Sharpe, "Illustrated Book Study"; "Digital Image Basics: Some Explanations and Descriptions of Terms," *DIMP*, www.ilrt.bris.ac.uk/mru/dimp/section3/basics_prt.html.

131. The Bandwidth Conservation Society maintains a Web site to assist developers interested in optimizing performance while maintaining an appropriate level of quality: The Bandwidth Conservation Society home page, www.infohiway.com/faster/homebcs.htm.

132. Stephenson and McClung, *Delivering Digital Images*, p. 72.

133. Cornell University, "Online Imaging Tutorial."

4 Establishing a Quality Control Program

Oya Y. Rieger

QUALITY CONTROL (QC) is an integral component of each stage of a digitization initiative to ensure that the results meet your predetermined standards. QC encompasses procedures and techniques to verify quality, accuracy, and consistency. Sometimes a distinction is drawn between quality control and quality review (or quality assurance). The former refers to the vendor or in-house inspection conducted during production. The latter refers to the inspection of final products by the project staff.

Although a quality control program is an essential component of both in-house and outsourced projects, the documentation required and timeline for developing these guidelines depend on how the project is implemented. In-house projects may have the luxury of defining quality requirements and procedures incrementally, building on each completed stage. For outsourced arrangements, however, it is crucial to articulate a comprehensive, prescriptive quality inspection program before an agreement is signed, due to the binding nature of contracts.

Implementing a QC program can be very time and labor intensive, requiring special skills and equipment. The award winners from the first round of the Library of Congress/Ameritech competition have emphasized that quality control took more time than they expected.[1] Obviously, you will save money initially by not instituting a QC program or by having a very scanty one.

However, when assessing the cost of a QC program it is important to factor in the cost of doing without and consider the effect of poor image quality on future projects, file longevity, and your institution's Web presence. Although the hardware, software, staff skills, and experience required for QC may seem overwhelming, even basic, simple techniques are better than no QC at all. Adopt a QC program that will suit your budget, technical infrastructure, staff qualifications, materials, and available time.

While this chapter presents a methodology that applies to evaluating various products of digital imaging initiatives, it focuses on the most important deliverable: digital images. Richard Marisa's sidebar covers quality control for metadata.

PREREQUISITES FOR QC

Identify Your Products and Goals

Since digital imaging initiatives involve several stages and parties, it is important to determine exactly what will be produced and when. First, identify the products of the digital imaging initiative. These products include but are not limited to master and derivative images, printouts, image databases, storage devices, and metadata products including converted text and marked-up files. The nature of the documents and the applicability of test results to the whole collection will help determine your procedures. For example, the Library of Congress inspects all Xerox PCX files because of the potential problems with diffuse dithered images of printed halftones. On the other hand, LC

1. The Library of Congress/Ameritech, "Lessons Learned: National Digital Library Competition," *Library of Congress*, memory.loc.gov/ammem/award/lessons/lessons.html.

inspects only 10% of their text-based TIFF images, since each batch is nearly uniform.[2]

2. Library of Congress, National Digital Library Program, "Quality Review of Document Images, Internal Training Guide" (September 1996), *NDLP Internal Documentation*, memory.loc. gov/ammem/award/docs/docimqr.html.

Your quality control methods also depend on your goals, ranging from faithful representation to removing a color cast that was unintentionally introduced during the photographic process. In the first case, you assess how well the image matches the appearance of the original document and relays

Metadata Quality Control

Richard Marisa
Manager, Electronic Printing and Publishing Initiatives
Academic Technology Services, Cornell Information Technologies

Metadata has a central role in processing, managing, accessing, and preserving digital image collections. Because of the crucial role it plays in the life cycle of image collections, metadata review should be an integral part of a quality control program. Metadata quality control cannot be done once at acquisition, but is a continuous process over the life of a collection. Plan and budget for this reality for any digital imaging initiative. Like image quality inspection, metadata quality control can be automatic or manual, objective or subjective, comprehensive or based on a sample.

Focus on these aspects of metadata:

▸ *Fidelity*: Transmitting, maintaining, or archiving metadata online can corrupt data. Many host operating systems, storage devices, and backup schemes contain mechanisms to detect and even correct data errors. However, collections that outlast a physical device or computer operating system should employ an independent mechanism for verifying data fidelity. One such mechanism is the MD5 message-digest algorithm placed in the public domain by RSA Security, Inc.[1] The algorithm takes a message of arbitrary length and produces a 128-bit "fingerprint" or "message digest." The fingerprint may be calculated before and after transmission and the two compared to test whether the data have been changed.

▸ *Form and validity*: Metadata schemes encoded in markup languages such as XML may be parsed to determine whether the data is well formed (correctly formatted) and/or valid with respect to a DTD (document type definition). Metadata encoded in proprietary schemes (e.g., Elsevier's EFFECT Exchange Format) is not supported by parsers and requires custom programming to algorithmically check for the equivalent of correct format and validity. Note that what constitutes validity may change in subtle ways as metadata specifications evolve. Be sure to consult and archive the relevant version of the metadata specification.

▸ *Accuracy of derived data*: The archive may contain data derived by manual keying or OCR. You can apply different metrics to determine an error rate for characters or words; it is important to identify what metric is most appropriate to the specific work. A mathematics text may contain myriad special symbols, but even a high accuracy rate may not be

useful if the symbols cannot be represented. That is, derived data may not be useful if syntax (typography) or semantics (in this example, mathematical intent) cannot be recognized. Only manual proofreading can provide a high-quality estimate of the error rate in OCR. Sampling can be used to estimate error rates over a collection of pages or documents. Therefore, correction of derived data such as OCR text is very expensive, but the accuracy of current systems is improving rapidly.

▸ *Correctness of data*: Ensuring the correctness of metadata (e.g., title, page number, author name, etc.) is a manual process. In assessing certain metadata, such as subject, you must rely on the experience of a cataloger. The values of some fields (e.g., rights management) may change over time and should be reviewed regularly.

▸ *Availability of linked resources*: As documents are increasingly digital, digital collections may incorporate references to material already online. For example, an art criticism journal may be composed primarily of links to images, each of which is published online elsewhere. The integrity of the virtual journal then wholly depends on the availability of the referenced materials.

▸ *Accuracy and completeness of components*: Access all the components (images, text, etc.) referred to by the metadata to confirm their presence. Where the metadata specifies the types of components (e.g., files, object, etc.), check to see that they are in fact of the specified type; for example, is a component really a TIFF image? Also confirm that the metadata describes all archived components.

▸ *Dynamic metadata*: Some metadata is generated dynamically and depends on the current server environment rather than the image collection itself. For example, an image collection may consist of high-resolution TIFF images, but server software may be able to convert these images to JPEG or PNG format for Web browsers. The possible dissemination formats are a feature of the server and may vary through the document's lifetime as the server environment evolves.

Note
1. R. Rivest, "The MD5 Message-Digest Algorithm" (April 1992), rfc.roxen.com/rfc/rfc1321.html.

its characteristics (colors, paper, etc.). In the latter case, you evaluate how well it matches the appearance of the original scene (rendering intent), rather than the photograph (exposure) at hand.

Agree on Your Standards

To measure quality, you need to clearly define it for each deliverable and establish acceptable and unacceptable baseline characteristics. First, develop shared terminology to facilitate a common understanding of expectations among staff, vendors, scholarly users, and funding agencies. If you will need paper copies of digital images to replace the text-based originals, for example, clearly state the required paper stock, paper size, toner, duplexing, printer performance, and printing resolution.[3]

Occasionally it is impossible to communicate your requirements in objective, quantitative terms. Some quality attributes can only be articulated in subjective terms, leaving room for misunderstandings and misinterpretations. In such cases, it may help to include visual samples to illustrate the points you are making.

If you set low standards, your quality control program actually needs to be more rigorous, since you leave little room for any additional errors. It may be better to set a higher standard and allow for greater headroom in your tolerance range.

Determine a Reference Point

A key question to ask is, What are we judging the images against? Identifying a reference point is not always straightforward. For example, if conversion is based on an intermediate transparency, the digital image is two generations away from the original. It has been copied to film (first generation), and that film has been scanned (second generation). At each step in this process, you can see a loss of quality in both the representation of color appearance and the rendering of fine detail. What should be the reference point in assessing this

image: the original document or the microfilm? Or, for assessing grayscale images of photo negatives, how do you judge the rendition of a negative as an image without seeing how it would have been developed?

Because there may be different quality standards for master and derivative images, you need to determine which version of the image to assess. If you base image quality control only on the quality of the master image, you can easily miss poor interpolation or dithering on access files. Likewise, assessing only derivatives may cause you to miss problems with the quality of the master files. Sometimes it is important to evaluate quality at each stage and for each product.

Understand the Limitations of Current Knowledge, Practice, and Technology

While many predict that more automated image evaluation tools will be available in the next several years, the current practical methods and tools for color and tone evaluation and printing are inadequate. In addition, evaluating quality on-screen is not yet as comfortable as printout-based evaluation. However, it is difficult to get good-quality hard copies of grayscale and color images, particularly for continuous-tone documents.

SETTING UP YOUR QC PROGRAM

Scope

You can apply your QC program to the whole image collection (archival and/or derivative images) or to a sample of images. For example, during Cornell's Making of America project, paper facsimiles produced from all the digital images were compared against the original pages. NARA, on the other hand, evaluates a random sample of the digital images (including the master files, the access files, and the thumbnail files): ten images or 10% of each batch of digital images, whichever is more. If more than 1% of the images in a batch are defective, the entire batch is returned to the service

---

3. For sample guidelines for defining image and metadata quality requirements for monographic and serial collections, see *RLG Guidelines for Creating a Request for Proposal for Digital Imaging*, www.rlg.org/preserv/RFPGuidelines.pdf and *RLG Model Request for Proposal (RFP) for Digital Imaging Services*, www.rlg.org/preserv/RLGModelRFP.pdf.

4. Steven Puglia and Barry Roginski, "NARA Guidelines for Digitizing Archival Materials for Electronic Access" (January 1998), *National Archives and Records Administration, The Electronic Access Project*, www.nara.gov/nara/vision/eap/eapspec.html.

Color 101: Introduction to Color Theory
Oya Y. Rieger

The human eye perceives color according to the wavelength of the light that reaches it.[1] The spectrum—the band of colors produced when sunlight passes through a prism—includes billions of colors, of which the human eye can perceive seven to ten million.[2] Color attributes are defined by the intensity of light and by its spectral content. For example, in bright daylight we see more color, contrast, and saturation, whereas at low light levels, objects are less colorful. Color appearance is affected not only by the light source and its properties, but by haze, viewing angle, and the surface characteristics of the object viewed, such as gloss. In addition, color perception depends on the observer's biological and psychological characteristics. Because every individual processes the signals differently, color perception can be highly subjective. For example, eyes are equipped with chromatic adaptation, which enables them to adjust to the overall color shifts produced by different light sources. The eye tends to cancel out overall color shifts using white as the strongest reference point. However, this tolerance will depend on the vision characteristics of the observer.[3]

Color models define the properties of the color spectrum in standard, quantitative terms. Although "color model," "color gamut," and "color space" are often used interchangeably, they refer to slightly different concepts. Color gamut is the total range of colors reproduced by a device. For example, a color is considered "out of gamut" if its value in one device's color model can not be mapped directly into another device's color model. Color space refers to a particular variant of a color model with a specific color gamut. For example, the RGB color model includes several color spaces, such as sRGB. The system palette (or CLUT, color look-up table) is the set of colors offered by a particular computer system.

Color Models
The most commonly used color models are RGB, CMYK, and CIE Lab.[4]

RGB
The RGB color model is made up of red, green, and blue. It is also called the additive color system, since it combines light to produce a range of colors: mixing two primary colors creates complementary colors. For example, red and green are superimposed to obtain yellow. Both scanners and monitors use the RGB color model. Monitors emit varying amounts and intensities of red, green, and blue light through phosphors to simulate a wide range of colors. In a 24-bit monitor, each color channel (RGB) is assigned 8 bits, so each monitor phosphor can be any of 256 (2^8) shades. This translates to an RGB color gamut representing 16.7 million colors.

Standard RGB (sRGB) was created in 1996 to manage color in operating systems and the Internet. Since then, several companies and consortia have adopted it as their default color space.[5] It targets nonprofessional users who find other more sophisticated methods of color management too challenging. It is based on a calibrated RGB color space that is well suited for

monitors, televisions, scanners, digital cameras, and printing systems. However, Bruce Fraser asserts that the sRGB standards are too conservative.[6] Many printable colors lie outside the gamut of sRGB, particularly in the cyans, blues, and greens. Consider this before using cameras and scanners that limit their output to sRGB at the start of the reproduction chain. This color space is rarely appropriate for archival images and is generally suggested only for access copies.

CMYK
The principle underlying CMYK, which is composed of cyan, magenta, yellow, and black, is that objects absorb certain wavelengths and reflect opposing wavelengths. For example, a yellow object absorbs blue and reflects yellow. Called a subtractive system, it uses colored pigments and dyes that filter light, taking color away from white light. Printing and photography are based on the subtractive color model.[7] The type and quality of colorants and the paper stock quality affect the CMYK color gamut.[8] For example, coated paper produces a wider range of colors than uncoated paper with a rough surface.

CIE XYZ, CIE LAB, and CIE LUV
CIE (Commission Internationale de l'Eclairage) is an international organization that establishes methods for measuring color. In 1931, CIE developed CIE XYZ, the first CIE color space, which is considered to be the foundation of colorimetry. Two additional CIE color spaces, CIE LAB (or L*a*b) and CIE LUV (or L*u*v), were released in 1978. CIE LAB represents subtractive color space and is primarily used for color print production. There are three sets of signals: L* is the lightness of an object, ranging from 0 (black) to 100 (white); a* is redness (positive a*) or greenness (negative a*); b* is yellowness (positive b*) or blueness (negative b*). CIE LUV is based on the additive color model and is used with color monitors. The CIE color model values are based on how the eye perceives color and were determined by testing human observers. RGB and CMYK color models are device dependent: colors produced depend on the display devices and their parameters. CIE-based color systems are device independent: theoretically, assuming output devices are calibrated, CIE colors should not differ from one device to another. It is the color model of choice for the International Color Consortium.

Color Gamuts
As shown in plate 3, the CIE color space describes the entire range of colors that the human eye can detect. Since the CIE gamut is much larger than the RGB and CMYK gamuts, it forms the basis for color management systems software. For example, color film produces a wide range of colors, including some a monitor cannot display. Printed images have a color gamut smaller than transparency film or monitors. In general, RGB-based systems such as monitors can create more light colors

CONTINUED ON PAGE 65

bureau for reinspection and correction. If less than 1% of the batch is defective, only the defective images are returned for correction.[4] The LC guidelines for scanning pictorial images suggest following a routine inspection for the first ten inspection batches. If no batch is rejected, you can loosen your inspection; if a batch is rejected, tighten it.

These technical reports contain procedures you can use to develop a scientific sampling plan for individual images:

▸ ANSI/ASQC Z1.9-1993 *Sampling Procedures and Tables for Inspection by Variables for Percent Nonconforming*

▸ ANSI/ASQC S2-1995 *Introduction to Attribute Sampling*

▸ ANSI/AIIM TR34-1996 *Sampling Procedures for Inspection by Attributes of Images in Electronic Image Management and Micrographic Systems*

CONTINUED FROM PAGE 64

than a subtractive system, which conversely can create more dark colors. Each color model excels at producing its own primary colors.

Hue, Saturation, and Brightness (HSB)

Color can be defined by three properties: hue, saturation, and brightness. Hue is determined by the light wavelength off an object—green, blue, yellow, etc. Saturation, also referred to as chroma, is the intensity or purity of the color. Brightness, also called lightness, describes the level of light and where the value fits in a white-to-black continuum. One of the best ways to understand these three attributes is to experiment with the HSB correction option of your image enhancement software. For example, open a color image and use the Adobe® Photoshop® software's Hue/Saturation option under Image/Adjust.

International Color Consortium

The International Color Consortium (ICC) was established in 1993 by eight industry vendors to create and encourage the standardization and evolution of open, vendor-neutral, cross-platform color management systems. Their ICC Profile Format is intended to represent color consistently across devices and platforms. ICC specifications, which are based on CIE LAB and CIE XYZ, describe the information required to ensure proper mapping of color profiles to an ICC-compliant application.[9] Device profiles provide color management systems with colorometric device characterization to convert color data between native-device color spaces and device-independent color spaces.

The ICC standard profile is widely available to hardware and software developers. The acceptance of the ICC format by operating system vendors allows end users to move profiles and images with embedded profiles between different operating systems transparently. For example, a printer manufacturer can create a single profile for multiple operating systems. Although embedding the ICC profile into each image seems like an excellent way to ensure that the color profile information is carried along the digitization chain, practical application of ICC profiles for imaging purposes is under debate.[10] In addition, at this point most of the current image formats do not support embedded ICC profiles.[11]

Notes

1. For a detailed discussion of human vision and its characteristics, see Howard E. Burdick, *Digital Imaging: Theory and Applications* (New York: McGraw Hill, 1997); GretagMacbeth, *Fundamentals of Color and Appearance* (New Windsor, NY: GretagMacbeth, 1997); *The Secrets of Color Management* (Randolf, MA: Agfa Educational Publishing, 1997).

2. The number of colors the human eye perceives is very debatable and one can find different figures. The numbers used here are based on GretagMacbeth, *Fundamentals of Color*, p. 1.9.

3. The Eastman Kodak tutorial on color offers an example of chromatic adaptation: "Chapter II, Color Theory," *Kodak Digital Learning Center*, www.kodak.com/US/en/digital/dlc/book3/chapter2/index.shtml.

4. Sybil Ihrig and Emil Ihrig, *Scanning the Professional Way* (Berkeley, CA: Osborne McGraw Hill, 1995), pp. 55–76, includes several visual examples to demonstrate the differences between different color spaces.

5. Michael Stokes, Matthew Anderson, Srinivasan Chandrasekar, Ricardo Motta, "A Standard Default Color Space for the Internet—sRGB," Version 1.10 (November 5, 1996) www.w3.org/Graphics/Color/sRGB; "Color Management in Microsoft® Windows® Operating Systems, An Overview of Microsoft Image Color Management Technology," White Paper (April 1997) *Microsoft*, www.microsoft.com/windows/platform/icmwp.htm.

6. Bruce Fraser, "Out of Gamut: Stop the sRGB Bandwagon," *MacWeek* (July 31, 1998), macweek.zdnet.com/1229/op_fraser.html.

7. High-end flatbed and drum scanners give the option to convert RGB-scanned images to CMYK color model. However, it is usually not practical to scan in CMYK. It is advantageous only if there are no plans for image enhancement, the image-enhancement software can edit CMYK images, or if the sole end product is a print (prepress application). See Ihrig and Ihrig, *Scanning the Professional Way*, p. 60.

8. Matching an original document's gamut to the inks offered by a printing system is especially challenging for older materials. For example, block printing in multiple colors from natural dyes and pigments dates back to the 15th century. These colors may be very difficult to match with today's manufactured dyes.

9. "Specification ICC.1:1998-09, File Format for Color Profiles," *ICC Specifications, International Color Consortium*, www.color.org/profiles.html.

10. Dan Margulis, "How Color Management Failed," *Electronic Publishing* (July 1999): pp. 33–34, 36, 38–42.

11. The ICC specification details the requirements and options for embedding device profiles within PICT, EPS, TIFF, JFIF, and GIF image files.

Methods

Depending on your quality requirements and deliverables, you can use several comparison methods:

▸ View the image at 100% (1:1) magnification onscreen without referring to the original document (compare it to the original if fidelity is the goal).

▸ Use grayscale and color targets to evaluate color both visually and by checking RGB (red, green, and blue) values for different steps.

▸ Use histograms and charts to evaluate signal-to-noise ratio, tonal distribution, spatial resolution, etc.

▸ Examine solely the printouts created from the image, or compare the printouts to the original document.

▸ Use image quality control reports generated by integrated image quality control software.[5]

Franziska Frey cautions against judging archival image quality based on viewing images on monitors and making adjustments to accommodate the characteristics of the viewing devices.[6] With system-integrated targets and associated software, you can define pictorial quality objectively without using monitors and printers, but until built-in tests are more available, it will be common practice to evaluate images by viewing them onscreen.[7] It is prudent, however, not to modify the archival

---

Image QC Checklist

Prerequisites for QC

▸ Identify Your Products and Goals

▸ Agree on Your Standards

▸ Determine a Reference Point

▸ Understand the Limitations of Current Knowledge, Practice, and Technology

Setting Up Your QC Program

▸ Identify Scope

▸ Determine Methods

▸ Control the QC Environment
  ▸ Hardware configuration
  ▸ Image-display software
  ▸ Monitor set-up
  ▸ Color quality control instruments and software
  ▸ Color management
  ▸ Viewing conditions
  ▸ Human characteristics

▸ Evaluate System Performance

▸ Codify Your Inspection Procedures

Assessing Image Quality

▸ Evaluate:
  ▸ Resolution
  ▸ Color and tone
  ▸ Overall appearance

---

copy based on this evaluation. Either readjust initial capture parameters to bring image quality to an acceptable level, or make corrections only on the access versions. Controlling the viewing environment is crucial for objective onscreen evaluation.

Controlling the QC Environment

The impact of image-display conditions on perceived image quality is often underestimated. Under improper conditions, even a very high-quality image may come across as unsatisfactory. For example, viewed on a monitor that cannot provide a full palette of colors, a 24-bit color image might look heavily posterized, losing the beauty and variety of the colors in the master image.

The factors in onscreen image quality inspection include:

---

5. ImageXpert™ is an automated image quality measurement system that can be used to evaluate digital capture systems. The software offers a variety of image quality assessment tools to quantify a range of image attributes, such as dot quality (including roundness and dot placement), line quality (e.g., edge raggedness, sharpness, and connectivity), halftone quality, etc. It also offers a color measurement system, allowing colorimetric and densitometric measurement. ImageXpert home page, www.imagexpert.com/home.hmx. Another example is MITRE's Image Quality Measure (IQM)© program, which directly measures image quality without relying on a specific pattern, a constant scene, or a reference: "Image Quality Evaluation," www.mitre.org/research/mtf.

6. Digital Imaging for Libraries and Archives Workshop (Scanning Photographs session), Cornell University Library, Ithaca, NY, October 1998.

7. For example, JJT, Inc. relies on visual, subjective inspection while evaluating the pictorial images created for the LC's National Digital Library Program: John R. Stoke, "Imaging Pictorial Collections at the Library of Congress," *RLG DigiNews* 3, no. 2 (April 15, 1999), www.rlg.org/preserv/diginews/diginews3-2.html.

- Hardware configuration
- Image-display software
- Monitor set-up
- Color quality control instruments and software
- Color management
- Viewing conditions
- Human characteristics

Hardware Configuration

Use the best configuration possible to optimize viewing. Some of the essential technical features include:

- Sufficient RAM for speedy retrieval, especially if the inspection is based on large file sizes. Peter Hirtle recommends that you multiply your largest file size by three to calculate the ideal amount of RAM (e.g., for a 12-MB file, 36 MB).

- A large monitor (e.g., 17" SVGA) set to 1,024 × 768 pixels, which is capable of displaying 24-bit color and supporting a 72-Hz refresh rate (1,024 × 768 pixels requires 3 MB VRAM).[8]

Read your monitor's product literature, because a monitor's color purity and brightness vary by as much as 25% from the center of the display to the corners. Some monitors require a video board to adjust the RGB color transfer functions for detailed color matching. You can usually accomplish this by downloading software into the existing board. Different manufacturing specifications, quality, and the age of a monitor also can affect consistent color display.[9]

Liquid crystal displays (LCDs) provide sharp, bright, flicker-free images, however they have not yet challenged cathode-ray tube (CRT) displays for image viewing. Regardless of recent significant improvements, most LCDs continue to rely on an analog connection that requires the original digital signal to be converted to analog at the graphics card and then back to digital in the monitor. This conversion can degrade image quality. Among other drawbacks, LCDs have a limited ability to display highlights and shadows, offer a narrow viewing angle, do not produce as wide a range of colors as CRTs, and are more expensive and more short-lived than CRTs.[10]

Image-Display Software

There are many software applications for viewing and manipulating images.[11] Select your image-display software according to the characteristics of the file format you use, and the kind of inspection you support (e.g., fast retrieval, creation of thumbnails for quick browsing, zoom-in feature, color measurement, etc.). For example, to evaluate Kodak™ Photo CD™ images, use Adobe® Photoshop® software with the Kodak Photo CD Acquire Module plug-in to ensure correct mapping of Photo CD colors.

Monitor Set-Up

Color management begins with regular monitor calibration: adjusting a monitor's color-conversion settings to a standard so that images look the same on a variety of monitors.[12] An image can look different on different monitors for various reasons, such as variances in the phosphors used to radiate the light. Monitor-calibration hardware and its accompanying software are ideal; however, if you

---

8. Refresh rate is the number of times per second an image on the screen is rewritten. Low refresh rates cause flicker, which is even more evident with large monitors. VRAM is the video memory to support colors at a given monitor resolution. To determine the required VRAM, multiply the number of pixels at a given screen resolution by the number of bits assigned to each pixel and divide the total by eight to convert bits to bytes. For example, to view a 24-bit image on a monitor set at 1,024 × 768 pixels, you need 2.36 MB VRAM.

9. A guidebook from Rheinner explains why imaging requires specialized display systems and outlines how to evaluate the features and functionality of such systems: *Display Subsystems for Production Imaging* (Hingham, MA: Rheinner, 1998).

10. Richard Entlich on developments in display technology, "FAQs," *RLG DigiNews* 3 no. 6 (December 15, 1999), www.rlg.org/preserv/diginews/diginews3-6.html.

11. Lists of image-viewing and enhancement software are available in several Web-based and print imaging publications, including "Products and Services Guide," *AIIM International*, www.aiimguide.com; "Imaging & Document Solutions Product Guide," *Imaging & Document Solutions*, ww2.infoxpress.com/mfidm/imaging/buyersguides/bg1; "Phillips Document Management Source Book," bizlib.com: *The Business Information Library*, www.biz-lib.com/ZPHDMS.html.

12. This assumes that the capture device has been calibrated prior to capture. Scanners are calibrated by comparing a scanned image to the reference target provided by the system's vendor, such as a proprietary target or a standard one like IT8. NARA guidelines provide information on how to calibrate scanners for different types of materials for both capture and viewing: Puglia and Roginski, "NARA Guidelines."

do not have access to these, you may want to use your application program's calibration tools. For example, Adobe Photoshop software includes a basic monitor-calibration tool, which eliminates color cast in your monitor display, ensures that your monitor's grays are as neutral as possible, and standardizes image display.

To calibrate a monitor, set the gamma and white point:

- Gamma is the monitor's light intensity. Default gamma values depend on the computer model and the operating system.[13] Set the gamma to 2.2.[14]

- White point (color temperature) is the color displayed when all three RGB phosphors are fully illuminated. Set the white point to cool white (5,000 kelvin).[15]

Although these settings are ideal for image quality evaluation, they may not assist users, who rarely view images in a dimmed, controlled environment. Corbis, for example, suggests a white point of 6,500 kelvin for viewing, taking into consideration users' typical viewing conditions. Eastman Kodak also recommends a white point of 6,500 kelvin for a television that displays images. Several institutions with Web-based digital collections are providing online adjustment tools and instructions to help their users optimize their viewing environment by evaluating and setting resolution, color palette, color balance, and brightness and contrast.[16]

In addition to calibrating your monitor:

---

13. For a detailed discussion of the importance of gamma in displaying images, see Howard E. Burdick, *Digital Imaging: Theory and Applications* (New York: McGraw Hill, 1997), p. 270–272. For information on gamma and gamma adjustment for various computer platforms, see Charles Poynton, "Frequently-Asked Questions About Gamma," home.inforamp.net/~poynton/GammaFAQ.html; "CGSD Service, Gamma Literature," *CGSD (Computer Graphics Systems Development Corporation)*, www.cgsd.com/papers/gamma.html; Portland Photographics home page www.portphoto.com (provides visual illustrations on gamma).

14. Some institutions using Apple Macintosh® computers prefer a gamma of 1.8.

15. The color of light sources is measured in kelvin (K). It describes how bluish or reddish a light source is. Noon daylight is around 5,000 K and an overcast sky is near 6,250 K.

16. For examples of online tools that help users optimize image quality, see "Monitor Adjustment Target," *National Archives and Records Administration, NARA Archival Information Locator (NAIL)*, www.nara.gov/nara/target.html, and "Setting the Resolution," *Photography Collection, Denver Public Library, Western History/Genealogy Department*, gowest.coalliance.org/calib.htm.

- Leave the computer on for at least half an hour to allow the monitor display to stabilize.

- Set the room lighting and then adjust the brightness and contrast controls on your monitor.

- Clean your screen once a week according to the manufacturer's instructions.

- Avoid turning the monitor on and off frequently.

Color Quality Control Instruments and Software

Color quality control instruments and software provide full spectral measurement and matching of colors from a wide range of objects, surfaces, and monitors. They assign numeric values to colors for consistent color analysis and are available for both reflective and transmissive media. For example, using a densitometer you can measure the density of specific regions of a photograph and compare those values to a digitized version. The most commonly used color instruments include:[17]

- *Monitor calibrators/optimizers,* which are used to ensure onscreen accuracy by setting white and black point, gamma, and color balance.

- *Densitometers,* which measure the amount of light transmitted through or reflected from an object. They measure the strength (density) of colors, and are primarily used in printing, microfilm inspection, and photography.

- *Colorimeters,* which separate light into its RGB components to determine colors' numeric values. The RGB values are usually converted into one of the CIE color spaces so that readings from different color spaces are easily compared. Colorimeters can measure the difference between colors and are commonly used to calibrate and characterize monitors and printers.

- *Spectrophotometers,* which measure the spectral-power distribution of light by comparing the amount of light shined onto an object with the amount the object reflects back. Merging the capabilities of densitometers and colorime-

---

17. For a detailed discussion and demonstration of various color measurement instruments, see X-Rite®, *The Color Guide and Glossary: Communication, Measurement, and Control for Digital Imaging and Graphic Arts* (1998), www.x-rite.com/Documents/Mktg/L11-029.pdf, and GretagMacbeth, *Fundamentals of Color and Appearance* (New Windsor, NY: GretagMacbeth, 1997).

ters, a spectrophotometer is the most precise and versatile color instrument.

Color control instruments vary widely in price from $400 to $5,000, depending on their functionality. They cannot replace human evaluation of color, but provide the observer with quantitative data to describe image quality.

Color Management

One of the main challenges in digitizing color documents is to maintain consistent color from image capture through display and printing—applications, monitors, and operating systems display colors differently. Color management systems (CMS) reconcile the different color gamuts of system components to provide color consistency from one output device to another. For example, using CMS, you can simulate onscreen the colors of the printout. Using CIE color spaces as a foundation, CMS are composed of a mix of hardware and software components:

- *Device profiles*, which describe a system's color capabilities, including its color gamut, color production method, and device operation modes.

- *Device calibration*, which determines what deviations have occurred in the device profile during use, and what action is required to bring the device back into adherence with the standard.[18]

- *Gamut mapping*, which translates color models. For example, converting from RGB to CMYK color is not an exact process because the color gamut of RGB is larger than CMYK. Colors that are out of gamut must be mapped to the next closest color.[19]

Operating systems and some imaging applications now include color management software.[20] However, despite the trend for more CMS support,

these modules have not yet gained much acceptance and suffer from compatibility problems. As James Reilly suggests, "it is important to approach the color management issue in small, manageable parts as overall color management through the chain is still a daunting task."[21]

Viewing Conditions

An important prerequisite for image evaluation is a controlled viewing environment. For example, changing the angle of the light can significantly affect your ability to read poor-quality information. The monitor and the original document require distinct viewing environments. The original is best viewed under bright light, and the monitor works best in a low-light environment. However, a low-light environment is not a dark room. Viewed in the dark, an image on a monitor appears to lack contrast.

For an ideal set-up:

- Keep the windows mostly shaded.

- Keep only half of the overhead lights on, preferably not those directly over the monitor.

- Have the monitor face a wall rather than windows, to eliminate reflections.

The monitor should be brighter than any other light source in the room. Ideally, separate the monitor and the viewing space for examination of the original. Give your eyes time to adjust when switching between the two environments. If you compare originals to digital images, a color-viewing booth is highly recommended.[22] In a viewing booth, tilt the document to eliminate reflections and glare from the lights in the booth. Set the lighting to Noon Sky Daylight at 5,000 kelvin (the ANSI standard). If you do not have a viewing booth, place the original document in a position that minimizes reflections and glare, preferably under a lamp that uses a natural-daylight fluorescent bulb.[23] If you do not have

18. "Information: Color Management," *Dunaway Products*, www.dunaway.com/Secties/Information/ColorManagement/ColorManagement.html.

19. Color management systems map colors based on different methods, including perceptive mapping, absolute colorimetric mapping, relative colorimetric mapping, and saturation mapping. For illustrations of how these different techniques are used in mapping, see *The Secrets of Color Management* (Randolf, MA: Agfa Educational Publishing, 1997), p. 11. This guide also provides examples of how CMS work in calibrating scanners and monitors (pp. 17, 21).

20. For example, both Apple ColorSync® and Windows® ICM incorporated color management into their operating systems.

21. Digital Imaging for Libraries and Archives Workshop (Scanning Photographs session), Cornell University Library, Ithaca, NY, July 1998.

22. Color-viewing booths, such as The Judge II from Gretag-Macbeth, provide metered daylight and balanced light levels. The American Society for Testing and Materials (ASTM) provides several standards for visual evaluation of color differences: ASTM home page, www.astm.org.

23. You can purchase natural-daylight fluorescent bulbs from lighting supply stores. Be prepared to specify length (usually 2 to 4 feet) and color temperature (D_{50} or daylight 5000 kelvin).

access to special color-balanced lamps and if the area has windows, evaluate color during the middle of the day for the best natural light. Use a light table to view originals in transmissive media such as slides or transparencies. Most light sources allow you to set the light source at a certain color temperature, such as 5,000 kelvin.

The surround is the visual field that encompasses the central viewing area; for example, it can include the background of the image, or the walls and furniture. It can influence color judgment significantly in two ways:

▸ Any light that strikes the surround is reflected onto the image, affecting its appearance.

▸ Eyes adapt to the surrounding color, biasing color judgment.

To minimize surround effect:

▸ Inspect images against a neutral gray background (both monitor display background and walls). Turn off any desktop pattern in your monitor.

▸ Wear neutral colors such as gray, black, or white, since they may reflect on the monitor.

Human Characteristics

Image quality assessment requires visual sophistication, especially for subjective evaluations. For color documents, an experienced and trained eye will easily catch the details that indicate quality. Particularly in the fine arts, quality judgments, as Michael Ester emphasizes, are in the eye of the beholder.[24] Because of this subjectivity, ideally, the same person should evaluate all images, using the same equipment and settings. Staff need training to communicate color appearance information effectively.

Some color vision deficiencies are linked to a defective, recessive gene on the x chromosome. Since females have two x chromosomes and males have one, the chance of color-deficient vision is 1 in 250 females, but 1 in 12 for males.[25] Even among expert viewers, differences in judgments due to normal variation in the human eye are not uncommon. Several color vision tests can be used to evaluate a viewer's vision.[26]

Evaluating System Performance

Whether conversion takes place in-house or is outsourced, evaluate and monitor your system performance to ensure consistency throughout the conversion process. The Association for Information and Image Management (AIIM) defines "system" as an organized collection of hardware, software, telecommunications, supplies, maintenance, people, training, and policies. ANSI/AIIM MS44-1988 (R1993), *Recommended Practice for Quality Control of Image Scanners*, provides guidelines to ensure the best possible scanned image quality and to confirm that a scanner is operating at consistently high levels over time.[27] Reilly and Frey recommend these system tests prior to conversion:[28]

▸ Assess *resolution* by measuring input and output resolutions. Two flatbed scanners with identical optical resolution may exhibit different resolving power and detail and edge reproduction.

▸ Evaluate *linearity* to ensure that the dynamic range of the original document is captured without distorting the tonal values. Linearity depends on the optics of the capture device and the electronics used in analog-to-digital converters.

24. Michael Ester, "Building Towards Image Resources and User Environments in the Visual Arts," in *Proceedings of the 1993 Electronic Imaging and the Visual Arts Conference: EVA '93*, ed. James Hemsley, Neil Sandford, and David Saunders (Aldershots, Hants: Brameur Ltd., 1993).

25. GretagMacbeth, *Fundamentals of Color*, p. 1.7.

26. For example, color vision can be evaluated with the Farnsworth-Munsell 100 Hue Test, which is available from GretagMacbeth: "Munsell Color Vision Tests," www.munsell.com/munvis.htm.

27. See the *RLG Guidelines for Request for Proposal* for an example of how to articulate vigorous quality control requirements to maintain consistent output. For another example of defining system requirements (although not fully implemented during the conversion stage), see Library of Congress, Contracts and Logistics Services, "The Library of Congress Requests Proposals For Digital Images of Pictorial Materials, Solicitation No. RFP97-9," memory.loc.gov/ammem/prpsal9/coverpag.html.

28. Franziska S. Frey and James M. Reilly, *Digital Imaging for Photographic Collections: Foundations for Technical Standards* (Rochester, NY: Image Permanence Institute, Rochester Institute of Technology, 1999), www.rit.edu/~661www1/sub_pages/page17.pdf. Robert G. Gann, *Desktop Scanners: Image Quality Evaluation* (Upper Saddle River, NJ: Prentice Hall, 1999) provides a comprehensive guide to measuring and evaluating several scanner attributes.

▸ Check for *flare,* which indicates stray light in the optical system. Test for flare especially when scanning transparencies with a large dynamic range.

▸ Assess *scanner noise*—the random variations in capture systems. The signal-to-noise test indicates how much electronic noise a scanner's components generate, sometimes evident in the final scan as speckles or granular patterns, especially in dark areas. Noise should not change unless dirt builds up in the system or the scanner operator alters the standard scanning procedures.[29]

▸ Check for *artifacts,* such as moiré (wavy patterns), banding (varying lightness and darkness caused by improper lighting), newton rings (circular impressions), etc.[30]

▸ Evaluate *color reproduction* by testing for precise overlapping of color channels. The color registration test indicates the alignment of a scanner's charge-coupled device (CCD) array's red, green, and blue planes that is necessary to produce clean edges between colors. On scanners that have poor registration, a close inspection shows a rainbow-like halo next to fine lines.[31]

As Don Williams describes in his sidebar on image-quality metrics, you can also evaluate capture devices using various resolution, color, and grayscale test targets.[32] Frey and Reilly emphasize that "targets are about the scanning system and not about collections."[33] Bear in mind that the tests based on these targets are aimed primarily at characterizing the scanning systems.

At this point, there are no off-the-shelf, integrated hardware/software solutions for capture system assessment. System evaluation continues to rely on both subjective visual inspection and objective measurements with various hardware and software tools. Ideally, image quality control tools to evaluate resolution, tone, and color should be integrated into the image capture or management software. Integrated capture and QC tools would also facilitate recording and maintaining QC data to support future activities, such as migration from one file format to another.[34]

Codifying Your Inspection Procedures

After identifying the products that will be inspected and agreeing on standards, tools, and methods, develop a set of procedures for your QC program:

▸ Who performs the inspection, how, and at what stages, and the extent of the inspection

▸ The instruments, hardware, and software you will use, and units of measurement

▸ How to reject unacceptable products

For in-house QC components, write up the inspection procedures as a short manual or as workforms.[35] Quality control data have long-term value, from supporting different stages of quality inspection to facilitating future manipulations and migrations. In addition to your procedures man-

29. Although some image-processing software can measure noise, currently none is practical. A standard for an objective noise test chart for electronic still-picture cameras is in draft ("ISO/WD 15739 Photography, Electronic Still Picture Cameras, Image Noise Measurements, Scanners"); see *PIMA/IT10: Electronic Still Picture Imaging,* www.pima.net/standards/it10a.htm.

30. For example, because the high-end digital scanning back used at the Museum of Modern Art is very sensitive to voltage and light fluctuations, electrical noise, and vibrations, the images were inspected to ensure that there were no resulting artifacts. In addition, system evaluation included checking for artifacts caused by the CCD sensor defects, which manifest themselves as scan lines, uneven exposure, dead pixels, and noise in shadows. Linda Serenson Colet, Kate Keller, and Erik Landsberg, "Digitizing Photographic Collections: A Case Study at the Museum of Modern Art, NY" (paper presented at the Electronic Imaging and the Visual Arts Conference, Paris, September 2, 1997).

31. "Color Scanners: How We Tested," *PC Magazine Online* (March 4, 1997), www.zdnet.com/pcmag/features/scanners/hwt.htm.

32. For an illustration of how a digital camera can be assessed

using a Macbeth Color Checker color rendition chart and a spectrophotometer, see Daniel Grotta and Sally Wiener Grotta, "Image Quality Tests: Digital Cameras" (February 10, 1998), *ZDNet Products,* www.zdnet.com/products/content/pcmg/1703/268566.html.

33. Frey and Reilly, *Digital Imaging for Photographic Collections.*

34. One of the key issues discussed at an April 1999 NISO/CLIR/RLG workshop was the need for the development of industry metrics to characterize and evaluate systems: "NISO/CLIR/RLG Technical Metadata Elements for Images Workshop" (April 18–19, 1999), *NISO (National Information Standards Organization),* www.niso.org/image.html.

35. For example, the National Digital Library Program uses such a QC manual to outline step-by-step procedures for inspecting image quality: LC, NDLP, "Quality Review of Document Images."

An Overview of Image-Quality Metrics
Don Williams
Image Scientist, Image Engineering and Simulation Lab
Image Science Division, Eastman Kodak Company

	Image-Quality Components	Quality Control
Tone Reproduction: *The translation of an object's light intensities into light intensities of visual displays for given viewing conditions or preferences.*	Determines how dark or light an image is, as well as its contrast. The most important image-quality metric because the effectiveness of all other image-quality metrics assumes that the tone reproduction is satisfactory.	For objective evaluation, a tone-reproduction curve relates the light intensity of points in a scene to the intensity of those same points in the reproduction. Each component in an imaging system contributes to the tone-reproduction curve by virtue of a characteristic curve that describes the input/output behavior of that component. For digital systems, the characteristic curve is called the opto-electronic conversion function (OECF). For the most part, color management systems (CMS) manage both tone reproduction and color reproduction. With a little effort, individual OECF curves are easily derived with image utilities such as Adobe® Photoshop®. Practical articles describe this for simple camera/printer combinations.[1]
Color Reproduction: *The rendering of an object's input colors (hue and saturation) as output colors under selected viewing conditions or preferences.*	Other terms associated with color reproduction—and used interchangeably—are white balance or neutral balance.	At a minimum, measure the extent to which gray or neutral areas in an object are maintained (i.e., balanced) as gray or neutrals. This is especially important when capturing documents under different lighting sources. A better objective metric for color reproduction that often assumes balanced neutrals is Average Delta E*, a single number that is calculated by comparing the input and output colors in a visually uniform color space called L*a*b*. Reproduce a color chart with known L*a*b* values, compare them to the reproduced L*a*b* values, and calculate the average difference. The validity of this metric depends on whether the colors used in the target are representative of items to be scanned. CMS typically optimize color reproduction by minimizing the Average Delta E*. This is transparent to the user.
Resolution: *The ability to capture or reproduce spatial details.*	While dot-per-inch or bar target readings serve as generic measures, they are misleading and do not correlate well with judged image quality. Resolution is best quantified by measuring the line spread function (LSF), the extent to which light spreads in the imaging process. The smaller the spread of light, the greater the resolution. Because interpretation is difficult, the line spread function is often transformed into the Modulation Transfer Function (MTF) for analysis.	Several software tools are available for measuring MTF. Two of them can be downloaded from the Technical Committee on Electronic Still Picture Imaging Web site (www.pima.net/standards/it10a.htm). Each assumes the use of a slanted-edge target. One is a plug-in for Photoshop on the Apple® Macintosh® while the other, a MathWorks MATLAB™ version, is suitable for cross-system use. Scanner targets for use with these tools are available from Applied Image (www.appliedimage.com).[2] Another MTF software utility for use with sine-wave targets can be found at www.mitre.org. Sine-wave targets can be purchased through Sine Patterns LLC (www.sinepatterns.com, under Products, MTF Test Patterns).

CONTINUED ON PAGE 73

CONTINUED FROM PAGE 72

	Image-Quality Components	Quality Control
Noise: *Point-to-point light-intensity fluctuations in an image that are not part of the original object.*	Most often associated with spatially random fluctuations as in film grain. In electronic imaging systems, noise is usually due to the sensor and associated electronics. Measured by calculating the standard deviation of pixel count values in an image of a uniformly lit gray patch. The greater the standard deviation, the greater the noise and usually the poorer the image quality.	Currently, draft standards recommend ways to measure noise levels of digital cameras that can also be applied to scanners (www.pima.net/standards/it10a.htm). Any image software that can calculate the standard deviation over a selected region can measure noise. An integrated package is ImageXpert™ (www.imagexpert.com). Be careful to measure only the noise element that is associated with the scanner. Applied blindly, many noise-measurement tools measure not only random scanner noise but also noise due to the target, nonuniform lighting, and other artifacts. The PIMA Web site recommends how to separate these different noise sources (www.pima.net/it10a.htm).
Artifacts: *Nonrandom or correlated light intensity fluctuations in an image that are visibly objectionable because of their magnitude.*	Examples are distortion, streaks, nonuniformity, dust, scratches, or oversharpening. While all of these appear in a captured image, they are not important until they become objectionable.	While many artifacts can be quantified, the way in which they correlate to perceived image quality is relatively unexplored. Judging them is best accomplished through experience. ImageXpert™ offers some tools that may automatically detect certain artifacts such as streaking or distortion. For a number of artifacts, the existing tools are sufficient for detection and quantification. However, only careful visual examination can detect many very subtle artifacts.

Notes

1. For example, see P. B. Gilman et al., "Some Simple Techniques for Improving the Print Quality of Digital Images from Kodak Electronic Cameras," in *International Conference on Digital Printing Technologies*, NIP 12 (Springfield, VA: IS&T, 1996), pp. 227–232.

2. For practical information on how to use the Applied Image Test Target for MTF analysis, see also Robert G. Gann, *Desktop Scanners: Image Quality Evaluation* (Upper Saddle River, NJ: Prentice Hall, 1999), pp. 183–199, 280–282.

ual, use a networked database or worksheet to record and share inspection data (QC metadata).

Questions to address when deciding how to reject unacceptable products include:

- Who pays for correcting unacceptable products?

- What is returned for correction: only the unsatisfactory images or the whole batch?

- What kinds of comments are included with the unacceptable products?

- What is the timeline for receiving the corrected versions?

- Is a second and subsequent round of QC needed?

- How will the corrected products be reintegrated?

ASSESSING IMAGE QUALITY

Resolution

Regardless of the nature of the original document, resolution is key to good image quality. For text-based documents, which have distinct edge representation, inspect legibility, tone rendition, completeness, darkness, contrast, sharpness, skew, and uniformity. For grayscale and color documents, especially pictorial images, bit-depth and dynamic range together with resolution determine the quality.[36]

The quality of detail rendering can be affected by such factors as the nature and complexity of the

36. There are also general factors such as image orientation, cropping, image enhancements, image processing, and scanner artifacts, etc.

Table 4.1. Commonly Used Resolution Targets[37]

Resolution Target[38]	Description
IEEE (Institute of Electrical and Electronic Engineers) Standard Facsimile Test Chart	‣ Originally designed as a fax machine test target; adopted by the AIIM standards committee for electronic imaging. ‣ Allows you to detect defects in images and evaluate the readability of text from a black-and-white original. ‣ Composed of high-resolution, high-contrast, black-and-white patterns. ‣ Produced photographically. ‣ Size: 8½" × 11"
AIIM Scanner Test Chart #2	‣ Provides a synopsis of the capture quality using a variety of font types and sizes, halftone wedges, character sets, and lines with different angels. ‣ Printed ink on paper. ‣ Size: 8½" × 11"
RIT (Rochester Institute of Technology) Alphanumeric Test Object	‣ Allows you to judge output in carefully measured increments and is commonly used to assess resolution limits for text-based, bitonal imaging. ‣ Consists of random arrays of block letters and numbers. ‣ Comes in three densities for crisp black-and-white documents, manuscripts, and older books. ‣ Size: 3" × 3"
PM-189 (A & P International) *See figure 4.1*	‣ Contains all the resolution patterns (except a continuous-tone grayscale) recommended by the ANSI/AIIM MS44-1988 (R1993) standard.[39] ‣ Produced photographically. ‣ Size: 8½" × 11"

document, detail and background color, document density, color contrast, and subtleties of shading. For example, it is often difficult to measure detail in photographs, which do not exhibit the same sharpness or crispness as graphic or textual materials. In continuous-tone documents, detail capture may be best judged in the highlights and shadows, which are more difficult to reproduce than the middle tones.

Resolution Targets

Resolution targets are used to measure detail reproduction, uniform capture of different sections and components of a document, image sharpness, legibility of different fonts, and print density. Resolution targets were originally developed for the micrographics and photographic industries. They do not consistently report how well a digital system resolves detail at various spatial frequencies, but rather characterize where it fails. The results may be inconclusive due to target density, variations in lighting, and the introduction of sampling errors during digitization. But, because we lack other practical tools, these easy-to-use targets continue to be instrumental in assessing legibility and consistency of capture devices, especially for bitonal conversion.

To evaluate your system based on these resolution targets, digitize the target the same way you do documents in your collection. Then examine

---

37. Some of these targets will be discontinued or redesigned in the near future. Upcoming changes include RIT's ceasing to distribute the Alphanumeric Test Object after current inventory is depleted (Barbara Giordano, RIT, memo, May 13, 1999) and the development of a new MS44 target that will combine AIIM Scanner Test Chart #2 and RIT Process Ink Gamut Chart (Dan Jaramillo, Kodak, e-mail to Oya Y. Rieger, July 12, 1999).

38. The IEEE and AIIM targets can be purchased from AIIM individually or as a set (Scanner Test Target Set) that includes an IEEE facsimile test chart, two RIT process ink gamut charts, and ten AIIM scanner targets: "Education, News, & Trends," *AIIM*

International, www.aiim.org/infoservices/index.html. The RIT Alphanumerical Test Object can be ordered by phone: (716) 475-2739. PM-189 is distributed by A & P International: A & P International home page, www.zimc.com/apintl. *Compilation of Test Targets for Document Imaging Systems* (ANSI/AIIM TR38-1996) provides a directory of the most commonly used test charts and test patterns used in document imaging applications, including a description of their primary applications.

39. Grayscale bar is excluded as it is almost impossible to maintain consistent grayscale levels and high resolution patterns on the same paper. The company provides a separate 37-step grayscale.

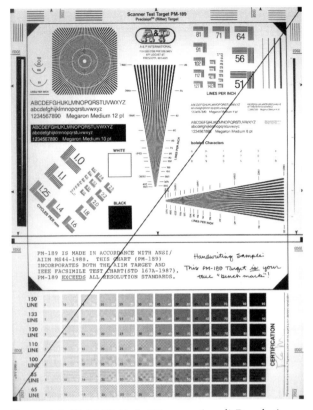

Figure 4.1 PM-189 (A & P International) Resolution Target

the image onscreen or printed out. To monitor system performance, you can save the image to compare to later images of the same target. Use the targets that represent the characteristics of your source documents. For example, if digitizing transparent media, use film-based targets; if your document is photographically produced, use targets containing photographic dyes. How often you use the targets depends on how much you scan, how often you calibrate your system, and how homogeneous your collection is. Don Avedon describes in detail how to use the resolution targets described in table 4.1, and discusses the pitfalls of sampling errors.[40]

Questions to Ask in Evaluating Resolution and Detail
Compare the digital images (or their printouts) to the original documents (or to the intermediates):

40. Don Avedon, *Quality Control of Electronic Images* (Silver Spring, MD: Association of Information and Image Management, 1997). Appendix E describes the resolution targets and problems caused by sampling errors.

Text/Line Art Documents

1. Is the stroke adequately reproduced?
2. Is the significant detail adequately reproduced?
3. Is the smallest text readable?
4. Are individual line widths (thick, medium, and thin) rendered faithfully?
5. Are serifs and fine detail rendered faithfully?
6. Are adjacent letters as separate as they should be?
7. Are the open regions of lowercase characters retained (i.e., not filled in)?
8. Are the edges of individual letters or shapes as smooth or well defined (not ragged) as the original?[41]
9. Is there good contrast or differentiation between the text and the background?
10. Is there even illumination across the image (i.e., is the image washed out or too dark)?
11. Is there a gray cast or streaking in the background?
12. Is the document fully reproduced?

Continuous-Tone and Halftone Documents
(The first three questions apply only to continuous-tone documents.)

1. Is the stroke adequately reproduced?
2. Is the significant detail adequately reproduced?
3. Is fine detail in the darkest and lightest portions retained?
4. Are there even gradations across the image (e.g., no banding, streaking, newton rings, or graininess)?
5. Is the image free of a moiré effect?[42]

41. Edge raggedness is the smoothness or straightness of edges along lines at very close inspection. Pay special attention to curved and diagonal lines on characters and line graphics.

42. Moiré patterns are most noticeable in the lighter regions of an image and in areas of "low activity" (e.g., in the sky portion of a landscape halftone). In portions with busy content (high activity), moiré is often hidden. Evaluate halftones onscreen only at 1:1 (100%); any other view might introduce distortions not native to the image file.

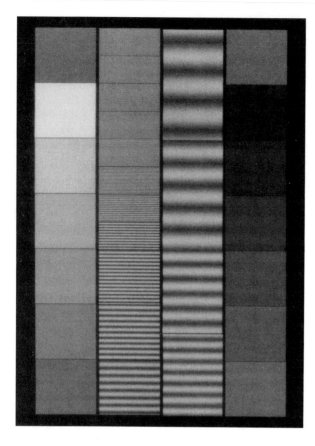

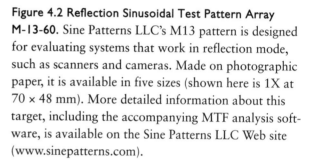

Figure 4.2 Reflection Sinusoidal Test Pattern Array M-13-60. Sine Patterns LLC's M13 pattern is designed for evaluating systems that work in reflection mode, such as scanners and cameras. Made on photographic paper, it is available in five sizes (shown here is 1X at 70 × 48 mm). More detailed information about this target, including the accompanying MTF analysis software, is available on the Sine Patterns LLC Web site (www.sinepatterns.com).

6. Is the significant informational content adequately reproduced?

7. Is the document fully reproduced?

8. Is the image too light or too dark?

The Promise of Modulation Transfer Function

As described in chapter 3, the Modulation Transfer Function (MTF) is a more reliable way to characterize how well detail is preserved across a broad spectrum of frequencies. It can also be used for periodic system calibrations. It is best suited for the characterization of grayscale or color systems. For information on new resolution standards in development, see the PIMA IT10, Technical Committee

on Electronic Still Picture Imaging Web site.[43] Figure 4.2 shows a test pattern array for MTF analysis.

Color and Tone

For color and grayscale images, color and tone complement resolution. For example, in scanned black-and-white photographs, high resolution and good detail capture will not compensate for poor tone representation. Photographic experts consider tonal reproduction most important in digitizing photographs; it is essential for details, especially in highlights and shadows. However, unlike image attributes such as detail and stroke, tone and color are not easily measured. Color and tone assessment may be highly subjective and change-

---

43. Draft standards promise significant contributions to standardized resolution evaluation: "ISO/FDIS 12233 Photography, Electronic Still Picture Cameras, Resolution Measurements" defines terminology, test charts, and test methods for performing resolution measurements for analog and digital electronic still picture cameras; "ISO 16067 Photography—Electronic Scanners for

Photographic Images—Spatial Resolution Measurements, Part 1: Scanners for Reflective Media" specifies methods for measuring and reporting the spatial resolution of electronic scanners for continuous tone photographic prints (see *PIMA/IT10*). There are currently no known standards for measuring visual resolution or the spatial frequency response (SFR) of scanners for print materials.

What Does the Colour Characterisation of Digital Still Cameras Standard Promise for the Digital Imaging Community?

Peter D. Burns
Image Scientist, Image Engineering and Simulation Lab, Image Science Division, Eastman Kodak Company

System color reproduction is often evaluated visually with displayed or printed digital images. Introducing a series of uniform neutral and color samples to the system enables a more quantitative evaluation. The captured or reproduced images of these test targets are compared to known colorimetric data for each target. This requires a transformation from digital image to input scene or object colorimetry. PIMA/IT10 has proposed a standard for color characterization of digital still cameras (DSCs), *Colour Characterisation of Digital Still Cameras Using Colour Targets and Spectral Illumination*, that is also likely to influence future methods for color document scanners.[1]

To understand the need for such standards, consider color management and how it applies to imaging for libraries and archives. Color management for imaging systems has historically included techniques for encoding and controlling how elements of a system store and adjust color information. The increased importance of open systems, not restricted to a limited number of input and output devices, has raised the importance of exchange between systems. To help this exchange, several standard file formats have been adopted. Today, for example, it is quite common to acquire a JPEG or TIFF image file and display the color image on a computer monitor. While this works well for many applications, it should be noted that neither file format contains any information (and therefore control) of the meaning of the resultant red, green, and blue color records. We do not know whether the image resulted from colorimetric image capture, which illuminant was used, or whether a video gamma look-up table has been applied.

A color management system requires the selection and communication of both a color encoding method and a metric.[2] The method defines the meaning of the image data. For example, this may require an approximate colorimetric specification at each image pixel by the set of three signal values. The color-encoding metric includes the color space and units used; for example, sRGB color space encoded over a specified range for each color record using 1,024 quantization levels each.

Proposed Solutions

The design of color imaging systems and ICC profiles requires careful characterization of each device. For digital cameras, a working group of PIMA has proposed several methods. The resulting color characterizations are expressed as transformations from camera data to a standard color metric. These transformations, when applied to camera image data, estimate scene colorimetry. The elements of the proposed standard include the means to:

▸ *Define a scene-analysis color space*. The spectral responses of the color analysis channels of DSCs do not, in general, match those of a typical observer, such as defined by the CIE standard colorimetric observer. Neither do the responses of different cameras necessarily match each other. To characterize digital cameras, therefore, it is necessary to take account of the spectral sensitivities, illumination, and reference color space.

▸ *Specify color targets*. The extent to which a color scanner's spectral sensitivities approximate those of human vision (as represented by the CIE color-matching function sensitivities) limits the flexibility of any device. For example, image data from a poor scanner may be transformed to X, Y, Z data for a given set of pigments or inks, but it fails for another set. Choose color test target samples, therefore, so that the colorants used (inks, pigments) are representative of those to be captured by the system. The GretagMacbeth™ ColorChecker® is a commonly used target that uses Munsell® paint colorants.[3] If a system scans photographic prints, whose colors are formed by three dyes, you can use a photographic target such as the Kodak Q-60 Color Input Target (IT8).

▸ *Establish consistent procedures*. The draft proposal discusses in detail the technical requirements and steps involved in developing the image transformations.

Many color imaging systems attempt to reduce or eliminate the colorimetric differences that result from different types of input material. For example, it is often important in digital prepress systems to assemble content from slide film and photographic prints for the same printed page. In digital library applications, however, it may be more important to acquire the image data that retains the unique characteristics of the originals, at least for the digital master file. Subsequently, distribution images can be modified for the intended output device. Future color management systems will be flexible enough to allow the selection of color rendering objectives and color exchange parameters. As part of a color management strategy, standards for color interchange and characterization offer tools for consistent image acquisition and compatibility with future systems for color imaging and archiving.

Notes

1. *Colour Characterisation of Digital Still Cameras (DSCs) Using Colour Targets and Spectral Illumination*, in "Program of Work," *PIMA/IT10: Electronic Still Picture Imaging*, www.pima.net/standards/it10/IT10_POW.htm.

2. E. J. Giorgianni and T. E. Madden, *Digital Color Management* (Reading, MA: Addison-Wesley, 1997).

3. C. S. McCamy, H. Marcus, and J.G. Davidson, "A Color Rendering Chart," *Journal of Applied Photographic Engineering* 2 (1976): pp. 95–99.

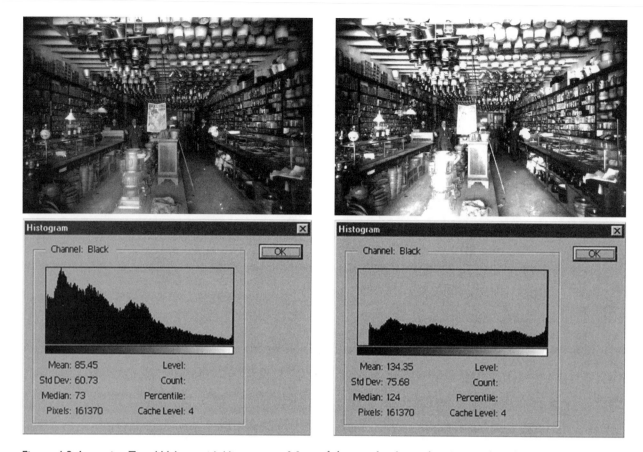

Figure 4.3 Assessing Tonal Values with Histograms. Most of the tonal values of an image that depicts a dark indoor scene will be gathered in the shadows (e.g., most of the values in the first image are in the 0–80 range). Blank ends at either end of the histogram may indicate that redistribution of the digital values is necessary to bring out highlights and details. For example, in the histogram of the image on the right, the values 0–15 have been clipped and the other values are fairly evenly distributed. The image has a washed-out look, increasing the contrast in the lighter areas and decreasing it in the darker areas.

able according to the viewing environment and characteristics of monitors and printers. Furthermore, color evaluation is challenging partly because it is relatively new to those involved in digital imaging initiatives and requires some basic understanding of color theory. In Cornell's Kodak Photo CD study, evaluators were more critical of resolution and detail in the images than color.[44] Illegible fine text or poor rendering of detail were more easily perceived and considered less acceptable than color shifts. Some study participants felt unqualified to judge color because it seemed to be

a daunting task and they had limited access to the required software and hardware. Columbia's Oversized Color Images Project concluded that color fidelity may not be judged essential in some digital representations, such as maps.[45]

To evaluate tone and color, you can compare the original document to a printout from its digital image, but it can be difficult to generate an acceptable color printout (especially for continuous-tone documents). To create evaluation printouts of continuous-tone documents, use a printer that can render continuous tones and regulate

44. Anne R. Kenney and Oya Y. Rieger, *Using Kodak Photo CD Technology for Preservation and Access* (Ithaca, NY: Cornell University Library, 1998); available from "Publications," *Cornell University Library Preservation & Conservation*, www.library. cornell.edu/preservation/pub.htm.

45. Janet Gertz, *Oversize Color Images Project, 1994–1995, A Report to the Commission on Preservation and Access* (Washington, DC: Commission on Preservation and Access, 1995), p. 8, www.columbia.edu/dlc/nysmb/reports/phase1.html.

tones and hues, such as a dye-sublimation printer. However, even with such a printer, color shifts due to gamut differences are usually inevitable.

Histograms

Evaluate tone reproduction to ensure linearity. Many image viewing and manipulation software programs allow the creation of histograms, which provide a graphical representation showing how tones are distributed throughout an image.[46] As shown in figure 4.3, the horizontal axis of a histogram displays the distribution of tones running from dark (0) to light (255). The vertical axis shows in relative terms how much of the image is assigned to each tone. A histogram that shows any clipping (compared to tonal values of the original document) in highlights or shadows may indicate a system's limited dynamic range—the loss of certain color values during capture or image editing. For example, the conversion of all tones lighter than a specified light gray level to white will lead to loss of detail.[47] The histogram also gives you a quick picture of the tonal range of the image, also known as the image key type (see chapter 3, page 37). Identifying the tonal range of the image helps determine the appropriate tonal corrections.[48]

The Library of Congress's RFP97-9 requires the measurement of tonal distribution to confirm that the system is capable of capturing images that have the full range of tones (i.e., the histogram needs to show continuity of sampling as specified with no clipping).[49] NARA's Electronic Access Project

(EAP), on the other hand, sets the maximum tonal range at the RGB levels for a color image from 8 to 247, since perfect linearity does not work well for online viewing (this range also leaves headroom for unsharp mask filtering).[50]

Color and Grayscale Targets

Color and grayscale targets provide some control over color and tone shifts. They can be used to check the photography, provide measurements for digitization, calibrate monitors and printers, and facilitate visual evaluation of digital images onscreen and printed out. For photo intermediates, they also facilitate matching to the *scene* rather than the *film*.[51] The European Commission's Open Information Interchange (OII) Standards and Specifications Web site provides information on standards that are used to interchange or record color information.[52]

Grayscale targets are instrumental in evaluating dynamic range. They can be assessed subjectively online (as an individual image or included with the digitized document) by evaluating the smooth gradations among the patches.

As described by Reilly and Frey, you can use grayscale targets to measure the linearity of the capturing device by comparing the density values of the print target to its digital version.[53] Plotting

---

46. For visual examples of how to evaluate and alter tonal curves and gamma values, see Sybil Ihrig and Emil Ihrig, *Scanning the Professional Way* (Berkeley, CA: Osborne McGraw Hill, 1995), p. 83.

47. Understanding the tonal qualities of the original can facilitate customization of tonal distribution by using the tonal curve options of your image-viewing software. For example, to bring out detail in highlights and midtones, you can adjust the curve to distribute more values to the higher digital range. However, this is achieved at the expense of a loss in the shadows, as there are a finite number of digital values to distribute from 0 to 255. Enhancing the midtones and highlights will compress the tonal values assigned for shadows.

48. For examples of histograms used to evaluate dynamic range, see James M. Reilly and Franziska S. Frey, "Recommendations for the Evaluation of Digital Images Produced from Photographic, Microphotographic, and Various Paper Formats," *Library of Congress, American Memory*, memory.loc.gov/ammem/ipirpt. html.

49. Library of Congress, "RFP97-9."

50. The NARA guidelines cover only the digitization requirements for online access (not preservation): Puglia and Roginski, "NARA Guidelines."

51. Michael Ester, *Digital Image Collections: Issues and Practices* (Washington, DC: Commission on Preservation and Access, 1996).

52. OII, "Colour Information Interchange Standards," *Info2000*, www2.echo.lu/oii/en/colour.html, includes information on the existing and upcoming color-related standards. CIE's Color Difference Evaluation in Images Technical Committee has drafted a technical report, *Methods to Derive Colour Differences for Images*, that aims to provide coherent recommendations for evaluating color differences, including the development of a standard set of reference images in a CIE color space, a sound statistical process for averaging color differences, a practical method to report color differences for images, a description of factors that affect the evaluation of color differences, and methods to derive color differences for digital images. CIE Technical Committee 8-02, "Colour Difference Evaluation in Images," www.colour.org/tc8-02. A new standard being developed by ISO seeks to define a scene analysis color space, a color target, metrology, and procedures for various situations: ISO/WD 17321 *Graphic Technology and Photography—Colour Characterisation of Digital Still Cameras (DSCs) Using Colour Targets and Spectral Illumination*. For more information about color-related standards, see *PIMA/IT10*.

53. Linearity evaluation based on grayscale target is described in Reilly and Frey, "Evaluation of Digital Images."

Table 4.2. Commonly Used Color and Grayscale Targets

Color/Grayscale Target [54]	Description
Kodak Color Separation Guide and Gray Scale (Q-13 and Q-14) See *plate 5*	• Includes both grayscale and color targets. • The grayscale has stepped, neutral values. It is composed of patches gradually shifting from white to black, produced on black-and-white photographic paper. • Color targets provide patches of representative colors for color appearance evaluation. • Comes in two dimensions: 8" (Q-13) and 14" (Q-14).
Kodak Q-60 Color Input Target (IT8) See *plate 4*	• Based on CIE LAB, is commonly used as a reference image for input calibration, and as a reference for subsequent changes to the images (including transfer to a new computer platform). • Manufactured in accordance with the ANSI/IT8.7/1 (transmissive) and IT8.7/2 (reflective) standards.
RIT Process Ink Gamut Chart	• Illustrates the full range of colors that can be printed using standard process inks (CMYK color model). • Shows how changes in printing variables affect the gamut of reproducible colors. • Useful for evaluating threshold settings of capture devices because even slight changes are easy to detect.
GretagMacbeth™ ColorChecker®	• Used in various applications from photography to scanner and monitor calibration. • Provides a checkerboard of 24 scientifically prepared solid-colored squares representing natural objects (e.g., human skin, foliage, and sky). • The color squares reflect light the same way in all parts of the visible spectrum so that they match colors under any illumination and any color reproduction process.
Reference Photo CD IMAGE PAC	• Created purely digitally, it represents the ideal neutral patches in PhotoYCC color space.[55] • Helps evaluate the dynamic range and color appearance for Photo CD images. • Can be used as a practical but crude tool to evaluate monitors.[56]

the density values for the grayscale steps against their digital values shows the performance of the system over a range of densities.

Using color targets to assess color appearance requires more sophisticated color measurement and evaluation skills. However, even an objective, visual comparison of the original color target to its digital counterpart may help you identify color shifts. Checking and modifying HSB (hue, saturation, and brightness) values using image manipulation software requires more advanced skills, but is fundamental to identifying problems with these aspects of color.

Cornell's Herbert F. Johnson Museum Online project, which aims to digitize more than 27,000 works, has relied heavily on both grayscale and color targets included with every image.[57] After visually inspecting the color target, staff adjust the hue, saturation, and brightness levels to diagnose and correct color-related problems. However, NARA guidelines argue that because color targets are printed on a printing press, they tend to be less accurate in color and density and to be less consistent from one copy of the color patches to another.

Some capture systems create a color look-up table (CLUT) to define the set of available colors by scanning a color chart and correcting subse-

54. Kodak targets are available from Eastman Kodak, 800-234-0426. To purchase the RIT target, call 716-475-2739 or order from AIIM International, www.aiim.org/infoservices/index.html, link to BookStore. A Reference Photo CD IMAGE PAC can be downloaded at "Fully Utilizing Photo CD Images, Typical Monitor Luminance for PhotoYCC Luma Values—Article Number 6," *Eastman Kodak Company*, www.kodak.com/US/en/digital/techInfo/pcd-102.shtml. Ordering information for GretagMacbeth ColorChecker can be found at the Munsell home page, www.munsell.com.

55. Kodak Photo CD images are based on Kodak's PhotoYCC color space. Similar to CIE Lab, it includes one channel of luminance (Y) and two channels of chrominance (C).

56. For instructions on how to use Kodak's Reference Photo CD Image to evaluate your image acquisition and monitor, see Kenney and Rieger, *Using Kodak Photo CD Technology*.

57. Cornell Institute for Digital Collections home page, CIDC.library.cornell.edu.

quent scans to color values from the chart. For example, Jenoptik's eyelike Digital Camera System uses GretagMacbeth™ ColorChecker® for this purpose. This feature can make QC and postprocessing more efficient and effective.[58]

Measuring RGB Values

For a more objective evaluation of color and tone, grayscale targets are commonly used to assess adherence to the ideal RGB (aimpoint) values set by a conversion project. You can compare the RGB values of each step of the grayscale on the target to standard values derived from the published reflectances of the target steps (see plate 5). Each patch of a grayscale or color target has a numeric digital value assigned to it. For example, for the perfect white patch, each RGB value should be close to 255, preferably without fluctuation in the three channels. NARA, in the Electronic Access Project, uses a Kodak grayscale target to measure aimpoint values based on the grayscale patches.[59] The NARA EAP guidelines give the ideal RGB values for different patches of the target (e.g., 105-105-105 at Patch M). The guidelines also specify the tolerance level for the numeric difference among the R, G, and B channels.[60] The closer the values are to each other, the more faithful the color representation. Even a minor difference in RGB values can give the image a slight but noticeable color cast (see plate 6).

Even for color documents, use both color and grayscale targets. Good color matching requires not only the information on the color scale, but also the grayscale control of shadows, midtones, and highlights. Don Brown of Eastman Kodak says "if you get your neutrals right, most of your color management work is done."[61] However, although these targets are very important, do not rely on them too much. As Reilly reminds us, "our job is

not to convert the grayscale bar but the document."[62] Reproducing the gray scale correctly does not guarantee optimal image quality; indeed, vigorous color corrections based on color bars may be dangerous. The resulting image can be perfect in terms of the grayscale, but may not be pleasing to the eye, as it may appear too flat.

Tips for Color and Grayscale Targets

▸ Digitize color and grayscale targets individually or with the document, depending on the nature of the documents and the capture process. Usually, color bars must not cover any portion of the document. Provide photographers and scanning technicians with detailed instructions on how to place targets to maintain consistency and to minimize wasted pixels. For an automated script to crop the control bars after digital images are created, place the color bars consistently.

▸ If you use a digital camera or rely on photo intermediates, the placement of the targets will depend on the aspect ratio of the document or the transparency. If the aspect ratio of the original document differs from that of the digital camera, you can place the control targets in the extra space. However, if the aspect ratios are the same or nearly so, there may not be any extra space for the control bars. Use the smallest available bar, even just a thin strip.

▸ Color and grayscale control bars fade over time and therefore need to be replaced periodically, depending on the frequency of use. Store them away from direct light, in folders or envelopes. You can check fading by comparing the target in use to a new, unused target. You may even consider cutting and saving a portion of the target for periodic comparisons. Or scan your new, unused target to compare to later scans from the same target.

▸ Record information about your targets (including sample scans and the original targets) as metadata to support future migration activities.

---

58. This information is based on an internal report by Andy Proft and Michael L. Bevans, Digital Photographers, from Herbert F. Johnson Museum, Cornell University.

59. Puglia and Roginski, "NARA Guidelines."

60. In Cornell's Kodak Photo CD study, the difference among the individual RGB channels was evaluated by dividing the smallest number of RGB reading by the largest number. For example, if the value for R is 95, G is 93, and B is 84, 84 divided by 95 equals 0.88. The closer this value is to one, the more faithful the color representation.

61. Kenney and Rieger, *Using Kodak Photo CD Technology*.

62. Digital Imaging for Libraries and Archives Workshop (Digitizing Photographs session), Cornell University Library, Ithaca, NY, July 1998.

Questions to Ask in Evaluating Color and Tone Appearance

Compare the image to the original document, an intermediate, or color/grayscale bars.

Grayscale Images

1. Evaluate tone appearance in the highlights (lighter sections), midtones, and shadows (darker sections). Are the details in these different sections captured without any loss?

2. Is the image too light or dark overall?

These questions are based on the grayscale targets used in photography and scanning:

3. How many grayscale bars can you count on your grayscale image?

4. If your grayscale target is numbered, at what numbers do you cease to discern distinct shades of white, gray, and black?

5. Is there an overall color shift on the grayscale target?

6. Use the information option of your image viewing software to read the color (RGB) values presented at different grayscale steps. How do they compare to the reference values provided by the grayscale bar? What is the difference between the smallest and largest values for each color channel for individual color patches (variance indicates color imbalance)?

7. Display a histogram of your grayscale bar image.[63] Are all digital levels from 0 to 255 used? Do you observe any clipping?

Color Images

1. Do you observe a color shift in the overall image?

2. Study the red, green, blue, and yellow colors. Do any show a color shift? Is it minimal or obvious?

3. Evaluate colors in the highlights, midtones, and shadows, especially red, green, blue, and yellow. Do any show a color shift? Is it minimal or obvious?

4. Is the image light or dark overall?

These questions are based on the grayscale and color targets used in photography and scanning:

5. How many grayscale bars can you count on your grayscale image?

6. If your grayscale target is numbered, at what numbers do you cease to discern distinct shades of white, gray, and black?

7. Do you notice an overall color shift on the grayscale or color target? If so, does it fall within your tolerance range?

8. As shown in plate 5, use the Window/Show Info option of the Adobe Photoshop software to read the color (RGB) values presented at different grayscale steps. How do they compare to the reference values on the grayscale bar? What is the difference between the smallest and largest values for each color channel for individual color patches?

9. Use the hue, saturation, and brightness adjustments of your image viewing software to evaluate the individual colors of the color bar. Comparing the colors on the color target to the original color target, is there a color shift to a certain color? Is it minimal or obvious?

10. Even if the color bar evaluation is satisfactory, compare different sections of the document to the image: is the color satisfactory?

Overall Evaluation

The final evaluation of an image should combine all the individual factors that contribute to its quality, such as capture system performance, resolution, dynamic range, and color accuracy. Your subjective evaluation should confirm your objective conclusion that an image is satisfactory. Remember that it is *impossible* to generate an image that will fully replicate the look and feel of a document. Ask these questions to evaluate overall image quality:

63. Rely on your image viewing software's user guide to find out how to create and evaluate histograms.

- Does the image convey all the significant information included in the original document (e.g., translucency in a watercolor painting, overlay in an oil painting, quality and texture of paper, etc.)? If not, how much does this affect your satisfaction with the image?

- Compared to the original document, is the image:
  - unacceptable?
  - adequate but diminished?
  - comparable?
  - improved?

- Will the user be satisfied with the image as a document surrogate, or will the image serve just as a basic access tool?

- Even if the image passes your subjective and objective inspection based on the grayscale and color targets:
  - Is the image's overall dynamic range adequate?
  - Are you satisfied with the general color appearance of the image?

5 Metadata:

PRINCIPLES, PRACTICES, AND CHALLENGES

Carl Lagoze
Digital Library Scientist
Computer Science Department, Cornell University

Sandra Payette
Digital Library Researcher
Computer Science Department, Cornell University

AS DIGITAL collections increase in size and complexity, their managers face the same issues that physical libraries confronted as they evolved into multifaceted research centers. How do users find the resources they need? What tools can help manage and organize these resources? How can the long-term preservation of these resources be ensured? How can access be limited appropriately? Metadata provides a critical part of the solutions to these questions and others.

Metadata is structured data about data—information that facilitates the management and use of other information. Two elements of this definition merit immediate note. First, "data about data" indicates the purely contextual distinction between data and metadata. For example, consider a movie and the review of that movie. The movie is data and the review is metadata (a rating) for the movie. Yet, the review is itself intellectual content—data—and in fact there may be other metadata for that review. Is the metadata for the review meta-metadata for the movie? Certainly, the way out of this quandary is to abandon the noncontextual distinction between metadata and data: "metadata" should be used only in the context of the relationship between individual resources.

Second, structure greatly enhances the utility of metadata. While certainly data can be about data even without structure, a key concern with networked information is the capability for machine processing, which was also the motivation for well-known metadata formats like MARC (MAchine-Readable Cataloging) and EAD (Encoded Archival Description). For the foreseeable future, automatic processing systems will depend on such structured

formats to effectively parse and disambiguate metadata.

Unfortunately, no silver bullet or package solution for metadata exists. Since so many metadata initiatives are evolving, the prudent course is to understand the current challenges, emerging principles, and best practices before implementing any particular metadata solution—and remain flexible.

THE CHALLENGE OF METADATA FOR NETWORKED INFORMATION

The practice of describing information is not unique to the digital age. However, the fantastic growth in the quantity, variety, and penetration of networked information resources on the Internet and the Web presents significant new challenges to the traditional methods of organizing and managing information.[1]

The Challenge of Scale

Traditional library cataloging has scaled up to institutions of great size, such as the British Library, the Library of Congress, and the Bibliothèque Nationale de France. However, over the past year the number of resources on the Web has begun to dwarf even the largest libraries. Estimates in mid-1999 put the total number of addressable

1. Some of the material in this section is based on a series of metadata workshops given at World Wide Web meetings and other venues. The authors acknowledge the contributions of Renato Ianella (Distributed Systems Technology Centre, Australia), Eric Miller (OCLC), and Stuart Weibel (OCLC) to these workshops and the concepts presented here.

Web pages on the order of five hundred million.[2] This scale presents considerable challenges to the economics of traditional library cataloging, which creates metadata records characterized by great precision, detail, and professional intervention. Some estimates figure each original library catalog record at $50–70. This high price is impractical in the context of the growth of networked resources and less expensive alternatives are needed for many of these resources.

The Challenge of Functionality

Traditional cataloging practices in libraries take a one-schema-fits-all approach using MARC/ AACR2. However, the many tasks that metadata must facilitate for networked information stretch this model to the breaking point. Automatic rights and access management stand out among the more complex requirements; others include content rating, system administration, and preservation. The variety of requirements suggests adopting a less monolithic, more extensible model for metadata.

The Challenge of Permanence

Networked resources are dramatically less stable than physical resources. The well-known problem of dangling URLs bedevils any librarian who tries to incorporate Web pages into a collection. Because there is essentially no cost in publishing networked content other than the fixed cost of establishing a server, there are no fixed boundaries between publishers and consumers of information, unlike the relatively stable boundaries of the print world. The lack of the traditional frameworks for management and stability makes for haphazard management and impermanent objects. Such a loose environment has a strong impact on the economics and incentive for metadata and points to the critical need for preservation-oriented metadata.

The Challenge of Integrity

The breakdown of traditional publishing roles has led to a crisis of information integrity on the Inter-

net. In the traditional print model, integrity was established both by the credibility of the publisher and the credibility of the information intermediary (e.g., library or bookstore) that made the material available. Because both can be circumvented on the Internet, it has become increasingly difficult to determine the integrity of information resources.

Questionable integrity of metadata can be non-malicious; for example, an author may assign a bad subject classification to a descriptive metadata record due to lack of training or care. It can also be malicious; in so-called index spamming, content creators seed metadata fields with misleading or incorrect information to affect how search engines rank their pages. An important challenge for networked information is to develop the mechanisms and policies to verify the origin of any information, including metadata.

The Challenge of Accommodating the Variety of Metadata

If traditional library cataloging practices are deficient in the context of networked digital resources, then what can replace them? Broad requirements for a replacement are:

- *Multiple metadata dimensions and forms*: A networked metadata framework must include individual components, or modules, each of which can be customized for a variety of content and use. An abbreviated list of metadata uses includes resource discovery, rights management, preservation, administration, provenance, and content rating.

- *Extensibility*: As the breadth of information on the Internet continues to expand and communities make creative use of digital information, they will continue to need new types of metadata or new uses of existing metadata. For example, adult content on the Internet has led to metadata that rates the appropriateness of content for various audiences.[3]

- *Community-specific management and creation*: Communities should maintain responsibility and leverage their own expertise in developing metadata modules. For example, legal experts

2. Steve Lawrence and C. G. Giles, "Search Engines Fall Short," *Science* 285, no. 5426 (1999): p. 295.

3. "Platform for Internet Content Selection," *World Wide Web Consortium*, www.w3.org/PICS.

should develop rights-management metadata, while information specialists administer resource-discovery metadata.

▸ *Functionality and simplicity*: Metadata for digital resources must be simple enough for easy creation and management. However, simplicity should not come at the cost of functionality.

The Warwick Framework defines a metadata architecture that conforms to this set of requirements.[4] The Framework offers an alternative to monolithic metadata architectures by specifying a modular architecture with "containers" for individual metadata "packages." This modularization encourages individual metadata communities— legal experts, librarians, and system administrators—to address their realm of expertise. Metadata communities can then define domain-focused semantics, registration, administration, access management, data authority, and storage and distribution.

The Challenge of Metadata Interoperability

The importance of interoperability springs from the nature of the Internet, which acts as a commons where multiple communities intermix and interact in nontraditional ways. By eliminating physical barriers, the Internet opens the possibility for spontaneous interactions among these formerly separate communities. In such interactions, one community's domain may use the work of another. For example, an art museum may point to a digitized city map from its Web page to provide directions for visitors. These spontaneous interactions underlie the need for common metadata standards that allow resource discovery, resource management, and content rating across traditional boundaries.

Metadata interoperability has three dimensions:

▸ *Semantic interoperability*: All metadata schemas express a classification of a semantic space into a set of categories. For example, the Dublin Core Element Set defines a set of 15 categories for resource discovery. Semantic interoperability

is the extent to which multiple instances of metadata express the same meaning in their categorization. A breach of semantic interoperability can arise when two communities use the same categorization scheme but have different interpretations of one or more categories.

▸ *Structural interoperability*: Each metadata record consists of a set of values assigned to the categories, or elements, defined by the specific metadata vocabulary. The structure of these values, or lack thereof, is relevant to the interoperability of the record. Humans are often quite capable of translating between unstructured values; many recognize that "Bill Gates" and Gates, William H." are the same person. On the other hand, US and British citizens interpret the date "10-1-99" differently. Computers are even worse at interpreting unstructured values so require strict definitions of structure.

▸ *Syntactic interoperability*: Creators of metadata want to store, exchange, and use metadata records from different sources. Such exchange requires common mechanisms for expressing metadata semantics and structure. "Syntaxes for Expressing Metadata" lists common mechanisms, including HyperText Markup Language (HTML), Resource Description Framework (RDF), and eXtensible Markup Language (XML).

FUNCTIONAL USES OF METADATA

Metadata for Resource Discovery

Collecting resources without giving people the means to locate them doesn't make a lot of sense. The daily frustrations of finding resources on the Web is the result of giving more attention to providing access to resources than to facilitating their discovery. Different types of resources may have different descriptive metadata associated with them, which may support particular approaches to discovery but not others. The Dublin Core Metadata Initiative is a leading effort to create a metadata standard that enables simple resource discovery across a wide range of digital resources.

Why Metadata for Resource Discovery?

Search engines have proliferated on the Web, exploiting over 30 years of information retrieval (IR) research to index full text. Millions of people

4. Carl Lagoze, Clifford A. Lynch, and Ron Daniel Jr., "The Warwick Framework: A Container Architecture for Aggregating Sets of Metadata," *Technical Report* no. TR96-1593 (Ithaca, NY: Cornell University Computer Science Department, 1996), cs-tr.cs.cornell.edu:80/Dienst/UI/1.0/Display/ncstrl.cornell/TR96-1593.

Syntaxes for Expressing Metadata

There are a number of ongoing initiatives to define mechanisms for delivering metadata over the Internet:

- *HTML META tags*: The ability to embed metadata within HTML documents has existed since at least HTML version 2.0. Simple attribute and value pairs can be placed in the HEAD section of an HTML document using the META tag. A browser does not display this information, but search engines can extract it. For example, an arbitrary metadata field called Title with the value "My Life Story" can be encoded as ‹meta name="Title" value="My Life Story"›. Current browsers and most HTML tools support META tag encoding; however, it is unwieldy for expression of more complex metadata.

- *XML*: Building on the lessons from the deployment of earlier technologies, XML provides the foundation for the increased dissemination and more powerful use of structured data on the Web.[5] Unlike HTML, XML is built on a well-developed model that cleanly distinguishes between describing documents structurally and specifying how a browser should display them. Unlike Standard Generalized Markup Language (SGML), it is not too complex and is more easily deployable with readily available tools.[6]

  Whereas HTML has a fixed set of markup tags, XML allows you to create and use new tags using DTDs (Document Type Definitions). At a minimum, an XML document must conform to basic syntactic rules that specify how to delimit and close tags. However, you can define new syntactic and structural rules that, for example, specify the hierarchical structure of tags (e.g., a tag specifying a chapter can only be used nested within a tag specifying a section) or the attributes that must be used with a tag (e.g., a tag specifying a chapter must include an attribute specifying the title of the chapter). Metadata developers can define new tagging syntax to express their metadata vocabularies.

  At the time of this writing, the use of XML for encoding metadata remains experimental. One notable project by the Computer Interchange of Museum Information (CIMI) resulted in a guide of best practices that includes a DTD for the basic 15 elements in the Dublin Core Element Set.[7]

- *Resource Description Framework (RDF)*: The World Wide Web Consortium (W3C) developed RDF to address many of the issues already described.[8] Numerous communities were actively developing metadata semantics and ad hoc mechanisms for expressing them. A mechanism was needed to allow community control, management, and access to domain- and function-specific metadata, yet permit interoperability among these separate efforts. Finally, it was time to expand the Web beyond simple links between documents to a network of semantically rich relationships that fully describe networked resources. RDF provides a data model to express complex metadata relationships, an XML-based transfer syntax for exchanging complex metadata, and a mechanism for machine processing of metadata.

  The combination of these components makes RDF a powerful and extensible mechanism for the expression, storage, and exchange of metadata semantics. Despite considerable experimentation with RDF in the W3C and research community, actual use by metadata practitioners has been relatively limited.

Given all of these alternatives for encoding metadata, implementers need to decide which is most practical. The power of both XML and RDF is compelling; however, both are in flux and supporting tools are in early release. On the other hand, HTML and XML (in its simpler manifestation with DTDs) are quite stable. Both offer the possibility of experimentation and deployment now. Furthermore, their relative simplicity encourages experimentation with more limited metadata approaches. Experience will show whether the costs of highly complex, detailed resource description will yield expected benefits. Formal evaluation of this and other questions is critical to the development of network metadata standards.

---

use these immensely handy search engines daily. However, their underlying technology has numerous inherent limitations:

- *Scalability*: Web search engines index information by downloading full content from sites. It is increasingly difficult to keep these indexes current; recent studies indicate that even the best search engines index only about 12–15% of Web content.[9] Even more problematic are the limitations of IR technology, which often lead to results of questionable value. The sheer size of the Web presents mathematical problems that

5. Tim Bray, Jean Paoli, and C. M. Sperberg-McQueen, ed., "Extensible Markup Language (XML) 1.0: W3C Recommendation 10 February 1998," *World Wide Web Consortium*, www.w3.org/TR/1998/REC-xml-19980210.

6. For information on Standard Generalized Markup Language (SGML), see "Overview of SGML Resources," *World Wide Web Consortium*, www.w3.org/MarkUp/SGML.

7. *Guide to Best Practice: Dublin Core* (Consortium for Computer Interchange of Museum Information, 1999), www.cimi.org/documents/meta_bestprac_final_0899.pdf.

8. Eric Miller, "An Introduction to the Resource Description Framework," *D-Lib Magazine* (May 1998), www.dlib.org/dlib/may98/miller/05miller.html.

9. Lawrence and Giles, "Search Engines Fall Short," p. 295.

reduce both precision and recall. In addition, the Web corpus is entirely nonspecific, which makes synonym clashes inevitable. Linguistic techniques can resolve some clashes, but these are imperfect and are the focus of ongoing debate within the information-retrieval community.

▸ *Intellectual property*: Access-control technology on the Web is relatively primitive, based mostly on domain-based access-control lists. Mechanisms that enforce the complex terms-and-conditions for access and use of intellectual content are still the subject of current research.[10] As a result, most content that is valuable to the rights holder is unavailable or kept behind special servers. For the short term, Web indexers can access and index much of the content on the Web. As Web-based access controls mature, the reach of indexers will be limited and, arguably, the full content of the most valuable objects on the Internet will be protected behind access-control barriers.

▸ *Format*: Existing information-retrieval-based search engines are limited to textual content. They index words in documents and process textual queries, returning lists of documents ranked by the appearance of the query words in the content. Applying this technology to images will require tools that handle queries such as "find images with cars in them" or locate images with features similar to those of another digitized image. Such tools are still in an experimental stage.

▸ *Context*: At a more abstract level, lack of context compromises the usefulness of indexing based on full content. The best tools to locate a resource are those tailored to the user's knowledge and context. For example, content-based searches of MEDLINE (the medical index at the National Library of Medicine) might be appropriate for a professional familiar with medical terminology. However, a high school student might not be familiar with these terms and would find content-based searching considerably less useful.

Structured metadata that can coexist with cur-

10. Mark Stefik, "Letting Loose the Light: Igniting Commerce in Electronic Publication," in *Internet Dreams*, ed. Mark Stefik (Cambridge, MA: MIT Press, 1996).

rent search-engine technology addresses these limitations. Metadata improves scalability, since full content does not need to be downloaded and indexed. Content providers may be more willing to freely distribute metadata for indexing, in lieu of the full content. Most significantly, metadata can include contextual information, leading to the possibility of developing community-customized indexes. For example, to facilitate searching MEDLINE by high school students, metadata might associate more common medical subject terms with the content.

Dublin Core Metadata for Resource Discovery

The Dublin Core Metadata Initiative (DCMI) developed in late 1994 out of the recognition of inadequate resource discovery on the Web. The DCMI produced a set of 15 descriptive metadata elements, the Dublin Core Element Set, to facilitate simple resource discovery. The elements are intended to be simple and intuitive, cross-disciplinary, international, and genre independent. The notion of simple resource discovery contrasts with domain-specific resource discovery, which may depend on characteristics of a resource familiar only to practitioners (e.g., esoteric topographic features familiar only to geospatial experts).

With the 15 elements firmly established, the DCMI currently focuses on the specification of qualifiers that add more detail to the descriptions and allow individual communities to express specialized semantics. Qualification of DC elements falls into two main categories:

▸ *Element qualifiers*, which semantically narrow the definition of a metadata class. For example, the qualifier "illustrator" for the element "creator" specifies a refinement of the semantics of the broader element.

▸ *Value qualifiers*, which aid in the interpretation of a value for a metadata element. One common example is a date scheme such as ISO 8601, which suggests that the string "1998-02-04" should be interpreted as February 4, 1998.

The whole notion of qualification threatens the interoperability of the DC Element Set—and any metadata set. If practitioners blithely define and use community-specific qualifiers that extend rather than refine element semantics, then Dublin Core descriptions using those qualifiers will not be meaningful to other communities that neither support nor understand them.

See the Dublin Core Web site for details about the initiative and the 15 elements and for latest details on completed work and work in progress.[11] At the time of writing, the most authoritative document on the Dublin Core is the Internet Engineering Task Force (IETF) definition of the elements.[12] Additional documents are under development to describe the encoding of Dublin Core in various syntaxes, including HTML, and to document the qualification process.

11. Dublin Core Metadata Initiative home page, purl.org/DC.
12. S. Weibel et al., *Dublin Core Metadata for Resource Discovery* (Internet Engineering Task Force, 1998), ftp://ftp.isi.edu/innotes/rec2413.txt.

The Dublin Core Element Set stands out as one of the most active and accepted metadata standards, used in more than 100 major projects in more than 20 countries. The element set, represented in HTML, can use existing technology and remains the best way to deploy resource discovery metadata in the current Web environment. In her sidebar, Dianne Hillmann compares MARC and Dublin Core and discusses the pros and cons of DC in comparison to MARC.

Metadata for Presentation and Navigation

Although Dublin Core enables discovery of a resource, it doesn't facilitate its use. Structural metadata is structured data that provides an orga-

---

Choices: MARC or Dublin Core?
Diane I. Hillmann
Head, Technical Services Support Unit, Cornell University Library

When considering whether to use MARC or Dublin Core (DC), remember that this is a question asked only in libraries.[1] MARC is a library-oriented metadata standard with very little overlap into the nonlibrary world. This is not to say that MARC is relevant only to libraries. Nevertheless, in the outside world where metadata is a much newer concept there is little impetus to adopt a standard like MARC, which though flexible and well supported clearly shows its bookish skeleton and requires specialized software. Libraries are used to coping with the directory structure, leader bytes, and complex interrelated fields and subfields of a MARC record, but its overhead is not necessarily a terrific selling point for expansion into other environments.

The DC, on the other hand, is not burdened by a legacy of 30 years of previous practice, much less the burden of millions of existing records that must still be managed. It is flexible, extensible for various objects and implementations, and interoperable with other metadata formats. That is the good news. The bad news is that it is also still evolving, still getting settled as a standard, still lacking documentation and users' experience. That is coming, but there is a certain risk to using any standard in its formative period.

What should a library do, given a choice? First, it is important to address these questions:

▸ *Who will create the metadata: catalogers, technicians, a machine?* Catalogers trained in MARC and Anglo-American Cataloging Rules (AACR2) tend to find working with DC frustrating. Where are the rules, the detailed documentation, and the rule interpretations? Those untutored in the traditions of cataloging find it much more comfortable, though they, too, need help maintaining consistency. Machines, delightfully literal as they are, require the least training, but their work often benefits from review by a trained eye.

▸ *Where will the metadata be stored: with images or other files, in a separate database, in more than one place?* Although many mappings already exist for "crosswalking" between metadata formats, specificity is often lost when moving between standards with different semantics or levels of complexity. A simple DC Contributor element, for instance, may contain a personal name, a corporate name, or a conference name. In MARC, these three forms are coded differently and coding also expresses how the name is formatted. In order to accommodate these kinds of unspecified names, MARC recently established a new field 720, Uncontrolled Name. Uncontrolled names may or may not even be in the order expected in MARC (last name, first name), but obviously sophisticated searching, sorting, or limitation based on name encoding cannot be promised for these sorts of names.

Mapping the more complex MARC data to Dublin Core also has its difficulties. Most existing MARC records include data that do not fit comfortably into DC and crosswalks map many of the fields that have no semantic equivalents into the DC Description element. This allows simple searching, but such data cannot be mapped back into MARC in any reasonable manner. Extended, or "qualified" Dublin Core using Extensible Markup Language (XML) and the Resource Description Framework (RDF) will allow links to many databases of name information, as well as subject, classification, and other standard vocabularies. When this capability is more widely used and understood, it should have a significant impact on the usefulness of crosswalks.

▸ *With whom will the metadata be shared: bibliographic utilities, exchange partners, other databases?* Most libraries already have many agreements and processes in place for sharing MARC data as part of membership obligations, for capturing

CONTINUED ON PAGE 90

CONTINUED FROM PAGE 89

routine cataloging data, and as part of special projects and contracts. In general, these agreements require that bibliographic records contain data at one of the standard "cataloging categories" or levels of quality. These levels are used to define expectations, to set financial charges, and to support detailed specifications for modification or augmentation of records. Dublin Core records at present do not support such expectations and may often need to be coded as the lowest possible MARC cataloging category. Most institutions have not yet addressed the implications for record sharing.

▸ *How will the metadata be maintained, particularly preservation information, names, URIs?* Strategies for maintaining metadata depend very much on where it is stored. Metadata stored separately from the described files also needs a maintenance strategy, since the data referred to may be subject to maintenance and refreshing that would affect the metadata. Metadata kept in more than one place (for instance, in a MARC database and some other file) will tend to diverge over time, and careful planning for linking and maintenance is essential to retain integrity.

In general, given the uncertainties and complexities of working with both an established and emerging metadata standard, it makes sense to be cautious. DC, though it may not work well at this stage as a storage format for traditional library data, still retains its attractiveness for libraries in two ways: as a lingua franca or lowest common denominator for searching metadata created under multiple standards, and as a cost-effective way to extend the reach of libraries into areas that have been traditionally underrepresented in library catalogs. Some of these areas include general Web resources, gray literature (preprints, reprints, theses, etc.), collections not prioritized for full archival treatment, curriculum materials, and other materials of current but not necessarily permanent interest. DC has the added advantage of having several simple tools readily available that can provide partial (or full, if you are not fussy) machine-generated metadata for Web resources, as well as easily templated options for simple metadata with the submission of materials. As a standard created with the Web in mind, DC is optimized for discovery and retrieval of digital objects on the Internet. This simpler metadata may not yet coexist seamlessly with more complex MARC records, but can be delivered in similar ways to users without necessarily the same investment in creation or commitment to permanence.

For the moment, most libraries will not want to answer the question, MARC or DC? unequivocally, but should use Dublin Core to extend their reach. The future will certainly change both our perception of the question and, most likely, the answer in ways we cannot yet see clearly.

Note

1. Formerly USMARC, now harmonized with CANMARC and UKMARC and formally known as MARC 21 (MAchine Readable Cataloging)

nizational blueprint and a means to group images together. Without this sort of structural glue, digital images are essentially just files. For example, a digitized book might consist of hundreds of TIFF images, but the roles and relationships of these images must be recorded so that they can be presented as a book. Structural metadata can support many functions including:

▸ Key access points (e.g., to the table of contents of the digitized book)

▸ Browsing (e.g., viewing of a structural map of the book)

▸ Navigation (e.g., turning the pages)

▸ Structural relationships (e.g., the parent–child relationship between sections and chapters)

▸ Presentation formats (e.g., different image resolutions)

The various approaches to recording structural metadata for digital resources and the collections to which they belong include two general categories: 1) using basic data-management techniques to store structural information about images, and 2) using document-encoding techniques to create document surrogates that mimic the structure of objects that digital images represent.

Creating Structure through Data-Management Techniques

Simple data-management techniques—storing and managing digital images—can result in metadata that facilitates access and use of digital resources. A very simple way to create structural metadata is to store images in file-system directory schemes that correspond to the hierarchical structure of the objects the images represent. For example, a directory hierarchy can mirror the structure of a journal, with digital images residing in named subdirectories that correspond to volumes, issues, and individual articles.[13] The directory scheme, itself, is implicit metadata that computer programs can interrogate to provide navigation of the journal.

Another approach to creating structural meta-

13. For specific examples of how to implement this strategy, see Carl Lagoze, "Metadata Creation: From Digital Images to a Database of Documents" in Anne R. Kenney and Stephen Chapman, *Digital Imaging for Libraries and Archives* (Ithaca, NY: Cornell University Library, 1996).

data is to record all structural information about images in a relational database. Applications can query the database to obtain structural information and retrieve images according to various roles (e.g., all the images that constitute a particular chapter in a book). This approach underlies many image management systems.

Creating Structure through Document Encoding

A more prevalent trend in digital imaging projects is the use of structured markup languages to bind images together and facilitate access to various structural components of objects and collections. The markup language approach can also be used in conjunction with data-management approaches by creating systems that transform data in underlying databases into a more standard representation (e.g., SGML or XML documents).

Encoding Document/Object Structure SGML has been used to encode a variety of electronic texts. One well-known effort is the Text Encoding Initiative (TEI), which has developed standard DTDs for encoding texts in the humanities.[14] Although the TEI has focused on the markup of intellectual structures within electronic texts (e.g., paragraphs, figures, verse), the TEI DTDs can also provide structured access to digital images. TEI has spawned offshoots such as the Ebind project, which specifically focused on digital imaging projects.[15] The Ebind team at the University of California at Berkeley developed an SGML DTD to bind together page images while preserving the structural hierarchy of the original document. For example, Ebind tags can be used to group images into logical subcomponents such as sections and chapters. Internationally, the Memory of the World program has adopted an SGML DTD as its structural metadata standard for the digitization of rare library materials.[16]

One disadvantage of SGML is that users must have SGML-aware software to view encoded documents. Not many software applications support SGML and the major browsers need special plug-ins. However, XML has recently emerged to provide many of the benefits of SGML without some of the obstacles. XML is poised to become a Web standard as major browsers are providing native XML support.

Significant digital imaging projects using XML include the Göttingen Digitization Center (GDZ) of the Lower Saxony State and University Library in Göttingen, which is relying on XML and RDF for its digitization program. The GDZ has developed a document management system that uses XML for standard data exchange.[17] Structured metadata for image documents is imported into and exported from an underlying SQL database in an XML format. The Making of America II (MOA II) project has developed an XML DTD to encode the structure of a range of archival objects, such as diaries, journals, letters, and photographs.[18] "The Making of America II DTD for Encoding Archival Objects" (following page) provides an example of how this DTD can be used to encode a diary, down to individual dated entries on a page.

How Much Structure? For digital imaging projects, the extent to which both the physical and intellectual structure of a work can be encoded depends on the format and granularity of the digitized data. For example, the physical structure of a book can be recorded by using SGML or XML tags to denote the different roles that page images play in the book's structure (e.g., title page, chapter, or list of references). However, when page images are the only data format available, tags can only assign the image files to structural groups (e.g., chapters)—tags cannot be used to mark any substructures within the text represented by an image. This leads to imprecision in some cases, such as when a chapter begins in the middle of a page. To define document structure at a more granular level, other digitized formats must supplement the page images to support lower-level access points. For example, ASCII-text versions of each page could support additional access points if division tags were placed around page sub-

14. Text Encoding Initiative home page, www-tei.uic.edu/orgs/tei.

15. "Digital Page Imaging and SGML: An Introduction to the Electronic Binding DTD (Ebind)," *Berkeley Digital Library SunSITE*, sunsite.berkeley.edu/Ebind.

16. For the United Nations Educational, Scientific, and Cultural Organization (UNESCO) standards announcement, see "Digitizing Library and Archives Collections," *WebWorld*, unesco.org/webworld/highlights/digitisation_0199.html. For information on the DTD developed by the National Library of the Czech Republic, see Memoriae Mundi Series Bohemica: Programme of Digital Access to Rare Documents home page, digit.nkp.cz.

17. Göttinger Digitalisierungs-Zentrum home page, www.sub.uni-goettingen.de/gdz.

18. "The Making of America II," sunsite.berkeley.edu/moa2.

The Making of America II DTD for Encoding Archival Objects

One of the major goals of the Making of America II project is to understand the role of metadata in facilitating the discovery, display, and navigation of various types of archival objects (e.g., diaries, manuscripts, photographs). The project team developed an XML DTD that is flexible enough to support the encoding of virtually any kind of archival object that is hierarchical in its structural nature. The DTD provides a model for aggregating digital image files and other data, recording key descriptive and administrative information about these files, and creating one or more structural views of the files that correspond to either the physical or logical structure of the primary source object.

A sample document from the project shows how the MOA II DTD supports these functions. The encoding of the diary of Patrick Breen (held at UC Berkeley) includes three basic sections:

- *File inventory* lists all of the files that compose the digitized version of the archival object (such as different image resolutions).

- *Administrative metadata* records technical information necessary for managing the digital archival object over time.

- *Structure map* encodes one or more structural views of the object. The example below shows the <div> tag to delineate the title page and the first two entries in the Breen diary. Each structural component can have multiple files associated with it, representing different versions of that entity. For example, both the title page and the diary pages exist in three image resolutions (three image files).

Also, diary entries can be viewed as text snippets from the separate SGML file that encodes the actual text of the entries on the pages.

```
<StructMap>
<div N='1' TYPE='Book' LABEL='Diary of
   Patrick Breen one of the Donner Party
   1846-57. Presented by Dr. Geo McKinstry
   to Bancroft Library'>
<fptr FILEID='HRJ1' MIMETYPE='image/jpeg' />
<fptr FILEID='LRJ1' MIMETYPE='image/jpeg' />
<fptr FILEID='LRG1' MIMETYPE='image/gif' />
<fptr FILEID='T1' MIMETYPE='text/sgml'
   TAGID='titlepage' />
<div N='1' TYPE='Entry' LABEL='Friday Nov.
   20th 1846 [Page 1]'>
<fptr FILEID='HRJ2' MIMETYPE='image/jpeg' />
<fptr FILEID='LRJ2' MIMETYPE='image/jpeg' />
<fptr FILEID='LRG2' MIMETYPE='image/gif' />
<fptr FILEID='T1' MIMETYPE='text/sgml'
   TAGID='entry1'/>
</div>
<div N='2' TYPE='Entry' LABEL='Entry sat.
   21st [Page 2]'>
<fptr FILEID='HRJ3' MIMETYPE='image/jpeg' />
<fptr FILEID='LRJ3' MIMETYPE='image/jpeg' />
<fptr FILEID='LRG3' MIMETYPE='image/gif' />
<fptr FILEID='T1' MIMETYPE='text/sgml'
   TAGID='entry2'/>
</div>
...
</div>
</StructMap>
```

Source: sunsite.berkeley.edu/moa2

---

structures (e.g., marking the exact beginning and end of a chapter).

Managers of digital imaging programs must determine the costs and benefits of investing in different amounts of structural metadata. Communities are developing guidelines and benchmarks to help with this decision making. For example, one of the goals of the MOA II project is to provide insight on how varying levels of metadata affect the functionality of encoded archival objects. Also, the TEI has created a helpful classification of different levels of structural encoding for digitized texts. In 1998, the Digital Library Federation (DLF) sponsored a meeting on TEI and XML in digital libraries that resulted in a set of draft guidelines for best encoding practices in libraries and a rationale for choosing from among five different levels of encoding, from a minimum that supports simple page-image linking and navigation

of simple structural hierarchies to more granular levels that unveil the rich semantic nuances of digitized documents.[19] Although these guidelines pertain to the use of the TEI Lite DTD (a subset of the TEI DTD), the report can be illuminating to those outside the TEI community.[20]

Encoding the Structure of Collections   While these efforts have focused primarily on providing structure to individual objects, others have focused on encoding the structural nature of entire collections

19. "TEI and XML in Digital Libraries" (two-day meeting, June 30–July 1, 1998, Library of Congress) www.hti.umich.edu/misc/ssp/workshops/teidlf. "TEI Text Encoding in Libraries: Draft Guidelines for Best Practices," version 1.0 (July 30, 1999), www.indiana.edu/~letrs/tei.

20. For more information about the TEI Lite DTD, see Lou Burnard and C. M. Sperberg-McQueen, "TEI Lite: An Introduction to Text Encoding for Interchange," Document no. TEI U 5 (June 1995), www-tei.uic.edu/orgs/tei/intros/teiu5.html.

of objects. Most well known in the library and archival world, the Encoded Archival Description (EAD) initiative has developed an SGML DTD as a standard for encoding finding aids—a variety of tools that both describe the organization of and provide access to archive and manuscript collections.[21] The EAD DTD is jointly maintained by the Library of Congress and the Society of American Archivists.

Although the EAD DTD is used to encode finding aids for archival collections, other DTDs are often used to encode the structure of digitized objects within the collections (such as correspondence or manuscripts). This continuum of structure has been one of the focal points of the MOA II project, which promotes a service model that integrates EAD-encoded finding aids with MOA II–encoded archival objects. The American Memory project of the Library of Congress also employs both the EAD for finding aids and its own TEI-compliant DTD for the encoding of books, manuscripts, pamphlets, and other historical texts.[22]

Metadata for Rights Management and Access Control

A major challenge for institutions that are managing digital collections is to develop strategies for protecting these materials once they are made accessible via digital library systems and the Internet. Solutions must be developed with an appreciation of the complexities inherent in:

▸ Protecting intellectual property rights

▸ Providing access controls for digital materials

▸ Facilitating transactions

▸ Protecting digital materials from malicious or unintentional attack

▸ Monitoring events in the life cycle of a digital resource

Traditional models for dealing with these problems are not easily replicated in a digital environment. To date, many techniques used to safeguard digital collections and protect rights are fairly coarse; they can be so general that they provide opportunities for abuse, or they can be so restrictive that they inadvertently deny access to authorized users. Digital libraries need well-conceived solutions for rights management and access control. Well-designed metadata is key to describing the complex policies that define rights, restrictions, and the rules that govern who can do what with digital resources. Furthermore, it is not enough to develop simple application-specific metadata sets. The problems of rights management and access control demand metadata models that 1) allow for the structured description of a wide range of complex policies, 2) facilitate the enforcement of these policies in highly automated environments, and 3) promote interoperability among different communities and application domains.

Background: Policies, Rights, and Enforcement

In discussions of access management, the term "policy" often arises. A policy is usually thought of as a formal articulation of the terms and conditions of an agreement, a set of access rules, or an expression of rights. For policies to be effective, they must be enforceable.

From the perspective of computer science, a security policy characterizes unacceptable behaviors.[23] For example, a policy can control access to resources by specifying the uses that are not allowed for a particular resource (e.g., students cannot delete the digitized book).[24] To be enforceable in digital library systems, however, such policies must be expressed precisely, so that computer programs and systems can understand them.

It has been argued that solutions for rights management and access control in digital libraries must accommodate both highly structured and explicit policies and the more ambiguous policies that arise

21. Library of Congress, Network Development & MARC Standards Office, *Encoded Archival Description Official Web Site*, lcweb.loc.gov/ead.

22. "American Memory DTD for Historical Documents," *Library of Congress, American Memory*, memory.loc.gov/ammem/amdtd.html.

23. Fred B. Schneider, "Enforceable Security Policies," *Technical Report* no. TR98-1664 (Ithaca, NY: Cornell University, Department of Computer Science, 1998), cs-tr.cs.cornell.edu:80/Dienst/UI/1.0/Display/ncstrl.cornell/TR99-1759.

24. One well-known model for this problem is an access-control matrix of subjects and objects and, at their intersection, a set of rights that define acceptable operations for particular subject/object combinations. Butler Lampson originally proposed the abstraction of an access-control matrix in 1971 in a paper presented at the 5th Symposium on Information Sciences and Systems. Also see Butler Lampson, "Protection," *Operating System Review* 8, no. 1 (January 1974): pp. 18–24.

out of intellectual property agreements and negotiations.[25] This has strong implications for the development of metadata models to support rights management and access control. In particular, metadata must be able to record key information about the agreements, obligations, and restrictions that apply to digital materials. A major challenge for digital library developers is to create metadata and encoding schemes that are well modeled, well structured, scalable, and interoperable. Ultimately, rights metadata must lend itself to automated and semiautomated policy enforcement in a highly distributed, heterogeneous environment consisting of many communities of users and providers and many types of digital resources and applications.

Major Initiatives in Rights Metadata

Various projects and initiatives approach the rights-metadata problem differently. The minimalist approach creates metadata element lists for managing rights and controlling access to resources in particular application environments. This approach often results in the creation of metadata that is customized to fit a particular community's needs and that may be limited in scope. Others have approached the rights and access-management question more abstractly and systematically. Models emerging from this broader approach will generally have more potential to scale in a complex networked environment consisting of many types of digital resources and multiple stakeholders. A number of significant initiatives are underway that have taken the systematic approach. Not surprisingly, some have come up with similar abstract models that tend to reflect the multidimensional nature of rights management and access control. Current activity focuses on three general areas: requirements definition, data modeling, and system implementation.

Requirements Definition   In 1999, the DLF sponsored a workshop to define requirements for enabling access in digital libraries.[26] Although the

workshop focused on requirements for access-management systems, it produced a conceptual model for rights-management metadata. Consistent with the basic access-control matrix, the DLF model determines authorization by the intersection of user roles, policy rules, and object attributes. The DLF model suggests that metadata about users, descriptive metadata about objects, and metadata about rights/duties must be considered in relation to each other. The report also poses several interesting challenges:

- Metadata models must account for the perspectives of three entities: publishers, intermediaries, and end users.

- Metadata that pertains to users and their privileges must not compromise privacy (e.g., an intermediary must not collect data that reveals everything a particular individual has read).

- Metadata schemes must accommodate the common ambiguity in intellectual property laws and concepts (such as fair use).

The DLF report unveils the multidimensional nature of access management and rights and, by implication, the need for rights metadata schemes that accurately model the problem domain.

Data Modeling   One of the most notable initiatives to take on the challenge of modeling rights management is the Interoperability of Data in E-Commerce Systems (<indecs>) project.[27] This effort was established to integrate projects from different sectors, including copyright societies, text and music publishers, and the recording industry. <indecs> has the support of international trade organizations, major publishing groups, and the International Digital Object Identifier (DOI) Foundation. The <indecs> operating perspective can be summarized as:

- Unique identifiers are critical to solving the rights management problem.

- Creations are complex entities with owners and users, plus a related array of events, rights, and transactions.

25. James R. Davis and Judith L. Klavans, "Workshop Report: The Technology of Terms and Conditions," *D-Lib Magazine* (June 1997), www.dlib.org/dlib/june97/06davis.html. Also see William Y. Arms, "Implementing Policies for Access Management," *D-Lib Magazine* (February 1998), www.dlib.org/dlib/february98/arms/02arms.html.

26. Caroline Arms, Judith Klavans, and Donald J. Waters, *Enabling Access in Digital Libraries: A Report on a Workshop on Access Management* (Washington DC: The Digital Library Federation, 1999).

27. <indecs> home page, www.indecs.org.

28. Godfrey Rust and Mark Bide, "The <indecs> Metadata

- Metadata is modular, but modules must connect easily and interoperate.

- Rights management and transactions must be highly automated to be effective in a digital environment.

The <indecs> metadata model clearly identifies the temporal and event-oriented dimensions of rights management.[28] This model can serve as a guide in the development of rights-management metadata for areas beyond e-commerce and publishing. The <indecs> project affords the important insight that descriptive metadata and rights-management metadata are inextricably connected and that considering one without the other can compromise the utility, interoperability, and effectiveness of both.[29] The <indecs> initiative hopes to provide a framework in which all metadata related to people, objects, and rights can be normalized to work together. Metadata interoperability has also been a major tenet of the Dublin Core Metadata Initiative and, accordingly, the <indecs> and DC communities have begun to work toward a common framework that will facilitate the integration of descriptive and rights metadata.[30]

System Implementation   Under the auspices of the International DOI Foundation (IDF), publishers and others have come together to devise a set of enabling technologies for identifying and protecting intellectual property in a digital environment. The DOI system will provide standards and technologies to support the creation, management, and use of identifiers and the requisite metadata for rights management.[31] To support the creation and resolution of unique identifiers on a global scale, the DOI employs the Handle System® developed by the Corporation for National Research Initiatives (CNRI). The next critical stage in the DOI

System is the development of rights metadata and the technical means for exchanging it.[32]

The IDF is committed to conforming to existing and emerging standards and has embraced the <indecs> data model as a foundation for its DOI metadata effort.[33] The DOI participants have established a metadata set of 14 elements that support the authentication and protection of intellectual property on the Internet. A subset of these elements forms what is called kernel metadata, the essential elements for interoperability. The DOI metadata initiative emphatically promotes the principle of well-formed metadata and the use of a well-conceived data model. Like <indecs>, the DOI group is banking on XML/RDF as the best way to express metadata and will be developing an RDF schema for the DOI element set. Both of these projects can be instructive for digital imaging projects that are grappling with complex design and implementation issues of rights metadata.

Metadata for Administration and Preservation

Data associated with the functions of administering digital resources, and ensuring their long-term preservation are often called technical and administrative metadata. They are sometimes also called preservation metadata, although preservation metadata can be thought of as a much broader category of data that potentially encompasses aspects of all the metadata functions described here. In some ways, the notion of preservation metadata is a bit of a paradox. In one context, it is all of the metadata necessary to preserve the integrity and functionality of a digital resource over time. Yet, preservation metadata must also be preserved. Thus, preservation metadata is subject to the same preservation challenges as the digital resource it supports.

While this problem may seem daunting, some practical steps can be taken toward a viable, long-term solution. A number of projects have identified data elements considered vital to preserving digital images over time. For example, RLG has come up with 16 metadata elements that describe key aspects of image files and the digitization process

29. Godfrey Rust, "Metadata: The Right Approach," *D-Lib Magazine* (July/August 1998), www.dlib.org/dlib/july98/rust/07rust.html.

30. David Bearman et al., "A Common Model to Support Interoperable Metadata. Progress Report on Reconciling Metadata Requirements from the Dublin Core and INDECS/DOI Communities," *D-Lib Magazine* 5, no. 1 (January 1999), www.dlib.org/dlib/january99/bearman/01bearman.html.

31. DOI: The Digital Object Identifier System home page, www.doi.org.

32. The Handle System® home page, www.handle.net.

33. Norman Paskin and Godfrey Rust, *The Digital Object Identifier Initiative: Metadata Implications*, DOI Discussion Paper 2, version 3 (February 1999), dx.doi.org/10.1000/131.

that created them.[34] The report of a 1999 workshop sponsored by the National Information Standards Organization (NISO), the Council on Library and Information Resources (CLIR), and RLG identifies functional requirements and preliminary metadata elements in three areas:

▸ Characteristics and features of images (e.g., MIME type, file format, spatial resolution, compression)

▸ Image production and reformatting features (e.g., image and system target information, scanner make/model/serial number)

▸ Image identification and integrity (e.g., universal identifiers, integrity checksums, important linkages to other metadata)[35]

Defining essential attributes of digital images and the processes and technologies that produced them is an important step in the development of metadata to support preservation strategies. It is critical that such attribute sets fit within well-designed metadata models that attend to the interdependent nature of various types of metadata. Several kinds of information will be important for preserving an object. Even a single digital image might require information about:

▸ Technical attributes (e.g., format, such as TIFF 5.0)

▸ Rights (e.g., the creator maintaining exclusive rights over access and use until 2021)

▸ Events that transpire (e.g., the publishing organization having dropped technical support)

▸ Critical linkages (e.g., a pointer to a Dublin Core record)

In their sidebar, Norbert Lossau and Frank Klaproth describe how the Göttingen Digitization Center uses the TIFF header fields to record and manage descriptive and technical metadata.

34. "RLG Working Group on Preservation Issues of Metadata: Final Report" (May 1998), *RLG*, www.rlg.org/preserv/presmeta.html.

35. David Bearman, "NISO/CLIR/RLG Technical Metadata Elements for Images Workshop, April 18–19, 1999," *NISO (National Information Standards Organization)*, www.niso.org/imagerpt.html; "NISO/CLIR/RLG Technical Metadata Elements for Images Workshop" (April 18–19, 1999), *NISO*, www.niso.org/image.html.

A Union Model of Preservation Metadata

Preservation metadata does not exist in isolation. Actually, it can be thought of as the union of all metadata that pertains to the continued discovery, use, and integrity of an object. Thinking about metadata for preservation in this way acknowledges the interdependent nature of the various types of metadata. Figure 5.1 assumes an underlying data model that addresses multiple functions (discovery, use, access control, preservation) and shows that different metadata views can be projected from such an integrated model. The preservation view inevitably encompasses the other metadata views. This is not meant to suggest that every aspect of every object must be preserved over time. Instead, it is a recognition that all types of metadata have the potential to affect long-term preservation.

A similar perspective on preservation comes from the electronic records community. Much attention has been given to record-keeping practices that ensure that evidence is preserved and much of this work has focused on capturing key information about transactions. With support

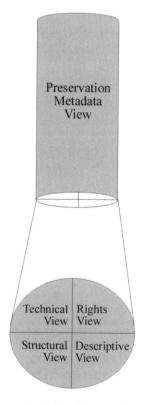

Figure 5.1 Preservation Metadata Model Projecting Multiple Views

TIFF Header:
A Reference Stamp for Image Files

Norbert Lossau
Head, Center for Digitization, State- and University Library Göttingen
Frank Klaproth
Head, IT Department, State- and University Library Göttingen

Tagged image file format (TIFF) has become the de facto standard for storing master and archival images.[1] It is a common output format for image capture software and therefore is often the first representation of the image for storage. A strength of the TIFF format is its file header option, which enables recording a variety of descriptive, administrative, and structural metadata within the file itself.[2] Other graphic file formats such as GIF or PNG also allow the entry of image-related information as file headers; however, the options are not as extensive and flexible. Since a number of the TIFF fields are not defined by the standard, the TIFF format is open and adaptable. TIFF may be potentially incompatible, however, if different institutions define these fields differently.

Why Use TIFF Headers?

In most file formats, when an image is saved, the scanning software creates about a dozen technical information tags automatically. Tags for information such as dimensions, compression, and color metric are required to enable graphic viewers to display baseline images. The addition of other header fields to this tag set is a common practice in several digital-conversion projects. For example, the Library of Congress requires fields such as DateTime and Artist in their TIFF headers. Recording metadata in TIFF file headers has several advantages:

- It ensures a long-term, close connection between document- and conversion-specific information and the image file, since the metadata is recorded directly in the file.

- TIFF headers contain structured information in the platform-independent ASCII format. This information serves as a "reference stamp" in the digital master file.

- Information from the TIFF header can facilitate the digitization workflow. For example, it is possible to restructure a digital book by reading out structure information from the TIFF headers.[3]

- Institutions can add other fields to describe the image and the scanning process. Certain "unlimited" fields allow the institution to freely structure and record content.

TIFF Header Content
at the Göttingen Digitization Center

In Germany, the report of the Technical Working Group on behalf of the German Research Council (Deutsche Forschungsgemeinschaft, DFG) recommends TIFF headers for ensuring the permanent identification of an image. Building on the experiences of other international digital imaging projects, the Göttingen Digitization Center (GDZ) defined a number of TIFF header fields and their structure for German digitization projects.[4]

Table A shows the default required fields for TIFF 6.0 images and the additional fields applied to a monograph by Edward Chappell entitled Voyage Of His Majesty's Ship Rosamond To Newfoundland And The Southern Coast Of Labrador. The information from the PICA/GBV Online Union Catalog, which serves the Göttingen State and University Library, can be directly cut and pasted into the TIFF tags.

Tools for Editing TIFF Headers

One of the main impediments to widespread adoption of TIFF headers is the scarcity of TIFF-header editing software. Commonly used image-viewing and manipulation programs, such as Adobe® Photoshop®, Paint Shop Pro, and Cerious® ThumbsPlus®, cannot create TIFF headers. Most can only read and display about 15 default fields of the TIFF file. Some programs, such as Graphic Workshop Professional™, can display more details; however, the display window cannot be scrolled and you can only read the part that is displayed. There are also special TIFF editors available; for example, Informatik Inc. offers Tiffkit™, a powerful suite of interactive programs and a collection of command-line-driven programs that allow you to edit single tags, rename files, and convert TIFF to GIF or JPEG.[5]

To support batch-mode processing, the German Research Council funded the development of software for the Göttingen Digitization Center. The software offers a template interface for DocumentName, ImageDescription, PageName, and Artist TIFF tags. This information can be added at the image level or recorded in a batch mode for every image in a group of files. The TIFF header can be created before or after scanning. The program builds on the official TIFF specification and allows the creation of TIFF headers even for externally produced TIFF files.

Image capture is often followed by image enhancement (e.g., despeckling, deskewing, etc.). To avoid overwriting TIFF-header information during the image-enhancement stage, create TIFF headers at the end of this process. In addition to recording data in TIFF headers, it is prudent to replicate some of the essential information in other locations. For example, at the Göttingen Digitization Center, the essential structuring and access information is also maintained in a platform-independent XML/RDF format.

CONTINUED ON PAGE 98

CONTINUED FROM PAGE 97

Table A. Sample TIFF Header Fields from GDZ

TIFF 6.0 Tags	Tag no.	Content
NewSubfileType	254	Default
ImageWidth	256	Default
ImageLength	257	Default
BitsPerSample	258	Default
Compression	259	Default
PhotometricInterpretation	262	Default
DocumentName	269	Catalog Id of the digitized document Sample: [PicaProductionNumber] PPN135661005
ImageDescription	270	|<DOC_TYPE>MONOGR|<HAUPTTITEL>Voyage of His Majesty's ship Rosamond to Newfoundland and the southern coast of Labrador|<AUTOREN/HERAUSGEBER>Chappell,Edward |<JAHR>1818|<ERSCHEINUNGSORT>London |<VERZ_STRCT>ChapVoya_13566100 Sample serial: Issue 5 in Volume 30 "|<DOC_TYPE>ZSCHR|Zentralblatt für Mathematik.| <JAHR>1945|<ERSCHEINUNGSORT>Berlin |<VERZ_STRCT>ZentMath_12345678_B0030_H005|" |
StripOffsets	273	Default
SamplesPerPixel	277	Default
RowsPerStrip	278	Default
StripByteCounts	279	Default
XResolution (dots per inch)	282	Default
YResolution (dots per inch)	283	Default
PlanarConfiguration (how the components of each pixel are stored)	284	Default
PageName	285	Physical page number (file name) Sample: Page 172 (r) 00000172 File name: Sample: 00000172.tif
ResolutionUnit	296	Default
(Scan)Software	305	Sample: SRZ Proscan incl. VersionsNR
DateTime (date and time scanned)	306	Sample: 19971020
Artist	315	Sample: Niedersächsische Staats- und Universitätsbibliothek Göttingen, Germany

Notes

1. For detailed information about the TIFF format, see Niles Ritter, "The Unofficial TIFF Home Page," home.earthlink.net/~ritter/tiff/#public.

2. For TIFF 6.0 specifications, see Adobe Developers Association, *TIFF™ Revision 6.0: Final—June 3, 1992* (Mountain View, CA: Adobe Systems Incorporated, 1992), partners.adobe.com/asn/developer/PDFS/TN/TIFF6.pdf.

3. Cornell's Digital to Microfilm Conversion project provides another example of using TIFF headers to support different processes. Given that file-size information for each image was recorded in the TIFF header and that all scanning was at 600-dpi resolution, a target noting the pixel dimensions and resolution was generated automatically from the TIFF header by the program for reel composition. Because variable reduction ratios were used during conversion, with this information one could then eas-

ily calculate the original page width and height. Anne R. Kenney, *Digital to Microfilm Conversion: A Demonstration Project, 1994–1996, Final Report* (Ithaca, NY: Cornell University Library, 1997); available from "Publications," *Cornell University Library, Preservation & Conservation*, www.library.cornell.edu/preservation/pub.htm.

4. For a description of the activities of the Göttingen Digitization Center, see Göttinger Digitalisierungs-Zentrum home page, www.sub.uni-goettingen.de/gdz; Norbert Lossau and Frank Klaproth, "Digitization Efforts at the Center for Retrospective Digitization, Göttingen University Library," *RLG DigiNews* 3, no. 1 (February 15, 1999), www.rlg.org/preserv/diginews/diginews3-1.html.

5. Informatik Inc. home page, www.informatik.com.

from the National Historical Publications and Records Commission, the University of Pittsburgh School of Information has identified functional requirements for record keeping that supports the preservation of evidence. The Pittsburgh project proposes a six-layer structure of metadata that would carry all the information necessary to allow a record to be used, including a unique identifier, resource discovery metadata, data structure, information on terms and conditions of use, and provenance information.[36] These requirements clearly show the interrelationship of multiple kinds of metadata in the preservation context.

Currently, two notable library-oriented projects are approaching the preservation problem with a well-conceived data model. Taking a systemic view of preservation metadata, the Cedars Project has synthesized metadata elements from several major initiatives (especially RLG's 16 elements and the Pittsburgh project) and has developed an extensive preservation metadata set that fits within the Open Archival Information System (OAIS) reference model for distributed archives.[37] Also, the PANDORA Project at the National Library of Australia has developed a logical data model that reflects an integrated view of the key functions of acquisition, management, use, and preservation of objects (specifically electronic publications).[38] These projects exemplify an approach that recognizes interrelationships between different kinds of metadata and the complexities inherent in creating metadata to support the long-term preservation of digital resources.

IMPLICATIONS FOR DIGITAL IMAGING PROGRAMS

As with other aspects of digital imaging, decisions about investing in metadata will be guided by a program's overall purpose and audience. Accordingly, program managers must evaluate the costs and benefits of creating and maintaining different kinds of metadata at different levels of granularity. This analysis must begin with a definition of functional requirements for metadata that includes the answers to several key questions:

▸ *How will users locate digital image objects?* Different descriptive metadata records may be needed to support discovery from online catalogs, custom application software, or the Web at large. Look to emerging standards such as Dublin Core whose ubiquity makes it a good choice for simple resource discovery. Also, identify potential conversions or crosswalks between the different metadata sets.

▸ *How will users interact with the digital image objects or collections?* Create structural metadata that meets the level of functionality an image object should provide to a user. At a minimum, design metadata that ensures reasonable presentation and basic navigation of image objects.

▸ *What policies are necessary to protect rights and provide access controls on objects?* Define access-control scenarios using a model such as an access-control matrix (subjects/objects/rights). Imaging projects must also consider the capabilities of existing security architectures when designing rights metadata.

▸ *How will the program assure permanence?* Define what permanence means for your institution and for particular digital collections. Identify what materials must be preserved, for how long, and under what conditions. Develop a plan for using persistent identifiers.[39] Carefully monitor emerging technical metadata requirements for images.

Note that any metadata created for one purpose

36. David Bearman and Ken Sochats, "Metadata Requirements for Evidence," www.lis.pitt.edu/~nhprc/BACartic.html; "Functional Requirements for Evidence in Recordkeeping," *University of Pittsburgh, School of Information Science,* www.lis.pitt.edu/~nhprc.

37. The Cedars Project home page, www.leeds.ac.uk/cedars. Lou Reich and Don Sawyer, *Reference Model for an Open Archival Information System* (OAIS) CCSDS 650.0-W-4.0, White Book, Issue 4, September 17, 1998 (Consultative Committee for Space Data Systems, 1998); see "ISO Archiving Standards—Reference Model Papers," ssdoo.gsfc.nasa.gov/nost/isoas/ref_model.html. Andy Stone and Michael Day, "Cedars Preservation Metadata Elements," Cedars Project Document AIW02, *The Cedars Project,* users.ox.ac.uk/~cedars/Papers/AIW02.html.

38. "PANDORA Project: Preserving and Accessing Networked DOcumentary Resources of Australia," *National Library of Australia,* www.nla.gov.au/pandora; "The Pandora Logical Data Model, Version 2" (November 10, 1997), *National Library of Australia,* www.nla.gov.au/pandora/ldmv2.html.

39. See Sandra Payette, "Persistent Identifiers in the Digital Terrain," *RLG DigiNews* 2, no. 2 (April 15, 1998), www.rlg.org/preserv/diginews/diginews22.html.

(resource discovery, navigation) has the potential to play a role in another function (e.g., access control, preservation). Good metadata should allow for flexibility, interoperability, and extensibility of functionality over time. The best strategies for achieving these goals are:

▸ *Data models*: Develop a well-conceived data model as a blueprint. If multiple metadata schemes are used, ensure that a framework allows them to coexist gracefully, and, ideally, complement each other. Data models should identify the relationships and interactions among multiple types of metadata.

▸ *Structure*: Well-structured metadata is important for machine readability. For example, XML provides a way to create structured, well-formed data that an increasing number of software applications can understand and consume.

▸ *Standards*: Look to emerging standards for expressing and exchanging metadata. XML is poised to become a standard transfer syntax for metadata on the Web and the next generation of Web browsers are slotted to have built-in XML capabilities.

▸ *Community semantics and rules*: To ensure consistency (e.g., in discovery, navigation, and presentation), collaborating communities will find it beneficial to embrace common semantics and rules for metadata creation and use. Flexible DTDs (for SGML and XML) illustrate this type of community standardization.

Well-designed metadata and conformance to emerging standards will keep digital imaging programs well positioned to use emerging tools, achieve interoperability, and migrate to new standards as they evolve.

6 Access to Digital Image Collections:

SYSTEM BUILDING AND IMAGE PROCESSING

John Price-Wilkin
Head, Digital Library Production Service
University of Michigan

THE CREATION of image files is never an end in itself. If we employ bitonal images as part of a program for preserving the contents of books and journals, our access strategies focus intensively on online (rather than print) use of the material. As we employ rich, tonal images to capture graphical information, we strive to support the use of those sources in new and more powerful ways. The use of image files to enhance access to library materials is now widely embraced as a powerful tool to extend the reach of library collections and perhaps even to minimize library costs through collaborative strategies.

As libraries build access systems with image files, decisions about image processing should be driven by costs, interoperability, user needs, and the way these shape libraries of the future.

- We must employ cost-effective strategies. We are working with large and significant institutional assets (i.e., our collections) and with finite capital resources. These collections are at the heart of our institutional missions, and making cost-effective use of them is no less imperative in the digital realm than in their physical state. Manual work to put these collections online is inadvisable, because the quantities of data are immense. As the size of these online collections grows, so too do the ramifications of decisions, and the prospect of being able to start over grows dimmer. It is important, then, to consider issues associated with image processing and long-term access at the point of digital conversion.

- As *libraries and archives* creating these Internet-based resources, we have a special incentive to integrate our work with that of our colleagues at other institutions. Profit is rarely at issue in our enterprise and a central motive is supporting the scholar in the most effective, thorough, and accurate way possible. As *cultural institutions* in this enterprise, our mutual aims and interests should lead us to create commonality across collections or institutions. Until we have well-defined methods for interoperability, the most effective way to support these aims is to ensure that we remain flexible about how we deliver information.

- We must take the user into account, which means employing diverse approaches for delivery. Flexibility and in particular the ability to offer images in multiple resolutions and multiple formats will improve discovery, navigation, and use by readers with varying levels of equipment, skill, and need. Flexible approaches also better position us to support future users, whose skills and technical resources will continue to improve. We must give special consideration to near-term uses, but our choice of strategies may also ensure that we are able to serve future users at low additional costs.

In some strategies, these themes—cost-effectiveness, interoperability, and the needs of the user—can converge in powerful ways. As we manage these assets, we must keep long-term goals in the forefront of our thinking. We must simultaneously advocate user needs, ensure that we are able to do *more* in the future *more effectively*, and make the best use of our collections and collection management dollars.

ITERATIVE SYSTEM BUILDING

All system building starts with an understanding of users and user needs. We can also build on the considerable work that has already been done, but must recognize that many of the model "solutions" are at best hypotheses that are yet to be tested. We must start by borrowing from (and acknowledging) other known successes and models and then developing an iterative process to improve what we have built.

Know Your Users

User- and use-centered design begins with a determination of the users' needs. The range of potential uses will probably be larger than the system can be made to accommodate, but an open systems strategy, which supports modular design and enables the introduction of other system components to support other needs, will ensure that the system can be expanded to serve a growing range of purposes. In building the University of Michigan's Making of America (MOA) system, system designers kept in mind a range that extended from popular pursuits (e.g., genealogical studies) to academic research. Quick searches and a variety of formats for viewing were implemented alongside persistent URLs and detailed information about the relevance of the results. The Early Canadiana Online (ECO) system took a similar strategy, plus accommodated both French and English language users with a choice of flattening or including diacritics.[1] The builders of the Museum Educational Site Licensing (MESL) system, constrained to offer the image collections only to faculty, staff, and students, took as their cues the primary instructional

needs of these more immediate constituencies.[2] A large body of literature covers user-centered design and needs assessment; our institutional missions and cross-institutional opportunities significantly shape our sense of user needs.

Inventory of Existing Models

The creation of access systems benefits greatly from a review of existing systems. In building Early Canadiana Online, project team members, each from a different institution, reviewed many full-text systems on the Internet. The members met to compare their findings, review successful and unsuccessful sites, and then assemble a list of features that drove both their digitization (e.g., full-text conversion and metadata strategies) and their strategy to build the access system. In building Michigan's MOA system, the Digital Library Production Service (DLPS) interface specialist drew not only from other Internet-based resources, but also from the organization's seven years of online system building with encoded text and image resources. MESL systems at the University of Virginia and the University of Michigan modeled significant pieces of functionality on instructional systems already deployed for art historical instruction. The Web is not yet as rich a development environment as proprietary and platform-specific environments, and the constraints faced by speaking to as wide an audience as the Internet itself makes the development and implementation process even more challenging. Nevertheless, Web browsers continue to gain power and flexibility, and the opportunity to borrow and reuse existing models will allow systems to grow in unprecedented ways.

Rapid Prototyping

Rapid deployment is facilitated by the wide availability of existing access systems. By pointing to existing systems, the ECO team was able to use URLs to cite complex examples of functionality in their system specifications. This method very effectively reduces communication problems between members of an implementation team and encourages the use of already available tools and services. Other areas of functionality can be defined by a prose description of the way that the proposed system differs from those implemented. Using a simi-

---

1. Early Canadiana Online is a full-text, Web-based collection of pre-1900 documents that were published in Canada or were published in other countries but written by Canadians or about Canada. The system includes some 500,000 pages digitized from microfilm by the Canadian Institute for Historical Microreproductions. *Early Canadiana Online*, www.canadiana.org.

2. *Images Online: Perspectives on the Museum Educational Site Licensing Project*, The Museum Educational Site Licensing Project, vol. 2, ed. Christie Stephenson and Patricia McClung (Los Angeles: The Getty Information Institute, 1998).

lar specification process, three generations of MOA have been implemented at the University of Michigan, each in a matter of a few months. It is also typical that a great deal of material is available before overall conversion is complete. In the ECO and MOA systems, staff began their implementation processes with a small proportion of the materials, while new material was being digitized and prepared for the online system.

Review

A speedy path from exploration to implementation not only serves immediate user needs but also allows testing on an existing system. The Canadian Institute for Historical Microreproductions (CHIM) implemented ECO rapidly, using known techniques for searching, viewing, and navigating the collection, and collected data on usage for subsequent revision of the online system. User responses not only validated a number of the decisions made by the original team, but also resulted in good suggestions for developing the system further. The first successful implementation of Michigan's MOA system was evaluated extensively by DLPS staff, in particular the interface specialist, and then by a group of students in Michigan's School of Information. Data and user comments were instrumental in designing the next phase of system development. The ability to evaluate a system's applicability to teaching and research, for example, by its *production* availability for courses and actual research, is a benefit that should not be underestimated.

Revision

Ultimately, the benefits of this development cycle must make their way back into the revision of the existing system. Making digital resources available is an ongoing expense, requiring appropriate commitments of staff and system resources.[3] While the ECO system has not yet been modified, the working system is an aid to the next specification phase, with comments and survey findings functioning as "exceptions" to the first system. Michigan's MOA

system has been revised twice since 1996, each time bringing the process outlined here full circle.[4] In this way, iteration has been a key to the MOA system's success in satisfying user needs.

It does not serve the user to delay the availability of online materials while seeking a perfect implementation. An iterative process best serves users, starting with an assessment of their needs, incorporating well-known strategies, and using the system itself for subsequent testing. The answers to the problems posed in serving users are as yet very incomplete and the array of questions for building access systems for images is far larger than for more familiar automated systems like online catalogs. Similarly, users' understanding of these systems is still limited; as they develop more confidence in online access to resources, users' demands will increase in number and complexity.

MASTER AND DERIVATIVE IN THE ONLINE SYSTEM

Libraries and archives must give careful consideration to the relationship of master and derivative in the online system—a consideration at least as important as iterative system building. As discussed in chapter 3, derivatives of image files are needed for more than building the online system. Bill Arms says powerfully and succinctly, "In the digital library, what you store is not what you get. The architecture must distinguish carefully between digital objects as they are created by the originator, digital objects stored in a repository, and digital objects as disseminated to a user."[5] Derivatives are likely to be tailored or even "tuned" for specific applications and thus will generally be easier for both humans and machines to use. In access systems, derivatives of image files can make it easier for users to pan and zoom in on large images, transfer a usable image to the desktop more efficiently, print an image more effec-

---

3. John Price-Wilkin, "Moving the Digital Library from 'Project' to 'Production'" (paper presented at DLW99 in Tsukuba, Japan, February 1999), jpw.umdl.umich.edu/pubs/japan-1999.html.

4. Maria Bonn, "Building a Digital Library: The Stories of the Making of America," www.umdl.umich.edu/dlps/mbonn-saunders. html.

5. Although not all digital library practitioners are as convinced that the master image should never be used as a format for delivery to the end user, Arms's statement is important food for thought about the relationship between master and derivative. William Y. Arms, "Key Concepts in the Architecture of the Digital Library," *D-Lib Magazine* (July 1995), www.dlib.org/dlib/July95/07arms.html.

Figure 6.1 Comparing 100 dpi derivatives created from 300 dpi 1-bit (left) and 300 dpi 8-bit (right) masters

tively, and even use an image without resorting to special plug-ins or helper applications. Moreover, for reasons such as security, rights protection, and valuable economic opportunities, it may not be desirable to allow end users direct access to master images. This need to use derivative formats of the image in building the access system raises a number of very important questions that influence the choice of formats.

One-to-Many:
Rich, Standard Masters Play a Pivotal Role

A rich master—one with significant inherent data—makes it easier to create any single derivative format, as well as making it possible to offer, simultaneously, many different kinds of derivative formats from the access system. Derivatives of image files are usually optimized for a specific purpose such as reading on the screen, high-quality printing, or extracting text for search and retrieval, among many others. The derivative's specialization is simultaneously the friend of the user and the enemy of longevity. While the master format, TIFF, is a community standard because of the amount of data it carries, its versatility, and its ability to be manipulated by an extraordinarily broad range of image-processing tools, the derivative requires of the user far fewer system resources, such as RAM or specialized software, and speeds viewing, printing, and other end-user applications. However, in optimizing the image for specific applications, derivative formats omit

information (e.g., through compression, reductions in resolution, or the omission of header information) or adopt proprietary formats (e.g., PDF). As a consequence, they are no longer as rich and cannot be processed with as wide a range of processing and management tools.

The richness of the digital master affects the quality and processibility of subsequent files derived from it. Michael Ester has summed up the effect on quality:

> When a digital image is reduced in resolution or dynamic range the difference in quality of a better capture process will be retained as the image is degraded. The notion that if a lower-resolution image is desired a poorer scanning method will do is true only with the caveat that the image will look much worse than necessary.[6]

Figure 6.1 shows the impact on image quality of derivatives created from various master files. Similar findings have also been observed in the effectiveness of content-based image retrieval indexing applied to images with a pronounced color cast. Ensuring quality in master files will result in better quality, greater accuracy, and/or lower-cost derivatives for enhanced access.

This dichotomy of flexibility and specialization becomes extremely important in the online system.

6. Michael Ester, *Digital Image Collections: Issues and Practice* (Washington, DC: Commission on Preservation and Access, 1996), p. 11.

The master file, with all of its richness and versatility, can be used to create many use formats. Creating derivatives from derivatives, on the other hand, results in degraded images or frustrating challenges. For example, creating one compressed format from another (e.g., JPEG from wavelet compression) results in compression artifacts (degradation) that *may* make the image unusable. Similarly, while an image in a proprietary format like PDF may contain all of the original information found in a bitonal high-resolution TIFF image, getting that information out (e.g., using the PDF to generate searchable text or GIF thumbnails) is at best problematic and is sometimes impossible. It may be possible to use the master image as *a part of* the access system, leveraging its versatility and richness. In many cases, this online integration of the master is either impossible or impractical; nevertheless, the likelihood of creating generations of access systems suggests the need to keep a clear understanding of the relationship between master and derivative. The creation of derivatives, essential in building access systems, benefits from an awareness of standards, an appreciation for the importance of investing in rich digital masters, and from integration of the derivative with a standard, rich master.

Fit the File to the Purpose

Using derivatives to create access systems requires balancing several factors, including the purpose of the derivatives and the unique characteristics of the master. The advantages and disadvantages of each image file format should be weighed in selecting the most appropriate formats for delivery (see table 3.4, page 52). A thorough investigation of the different formats may be a necessary first step. A significant body of literature, much of it on the Internet, covers the advantages and disadvantages of different formats for different purposes.

Just as image formats are optimized for different purposes, so too are the tools for processing images. The challenging process of choosing between image formats is often complicated by the selection of tools. For example, one tool can make creating a derivative expensive and thus impractical, while another tool may entail minor costs, making the choice of a given derivative format less consequential. For example, Adobe® PhotoShop® offers power and utility while simpler imaging applications such as ACDSee32™ open image files far more quickly but with a much narrower range of capabilities.[7] This dichotomy is no less true for tools used in creating derivatives. Programs such as tif2gif and ImageMagick offer a mix of often mutually exclusive capabilities that require consideration of the way they are involved in building access systems.[8] An inappropriate choice may deliver poorer information to the user or use resources less effectively in building the system. Test and experiment with different tools and formats to help match the format to the purpose.

Criteria

As we build digital libraries, and particularly Internet-based collections, a major component of delivery is display on a computer. The need to be able to display images in such a broad variety of user environments—with varying video equipment, operating systems, gamma settings, and even with different user interests—drives many of the criteria for selecting image file formats. Some criteria to consider include:

▸ *Proprietary versus open formats*: Many proprietary formats are supported only by a narrow range of applications (typically only those produced by the owner of the format) and thus their future is tied to the viability of those applications. The Internet is littered with "good ideas," particularly in the form of impressive plug-ins or helper applications with frighteningly short life spans. Ask whether the importance of the functionality in a format is so critical as to possibly limit how long you can offer images in that format. Instead, the user's ability to view the image without special supporting applications may be far more critical than a specific functionality. Just as most proprietary formats are tied to a narrow range of applications, most open formats are widely supported, or at least supported by a small number of commonly used applications.

---

7. ACD Systems home page, www.acdsee.com.
8. "Tif2gif," kalex.engin.umich.edu/tif2gif. "ImageMagick," www.wizards.dupont.com/cristy/ImageMagick.html.

▸ *Support for variant views*: Versatility is a key feature for access systems. Some formats are well suited to being viewed at different resolutions or in different segments. Adobe Acrobat® Reader, for example, can zoom in on an image for increasing levels of detail. This type of functionality will grow as imaging applications develop in intelligence and as the user's desktop environment supports greater levels of sophistication.

▸ *Cost-effectiveness*: Cost-effectiveness is key to scaling large collections, but it should be emphasized that an approach that is cost-effective in the short term may be costly in the long term. In shortsighted image capture projects, too little resolution and too much compression have been applied to save money on storage. Similarly, access systems should take both long-term and short-term costs into account. For example, an approach that is more disk intensive (and thus more costly in the short term) obviates medium-term or even long-term reprocessing of the image data. Or an approach that involves investing more up front in reliable automatic processing routines to remove costly manual processing may reduce near- and medium-term reprocessing costs. Ensuring that image files are versatile and can be easily processed is a key factor in cost-effectiveness; choose tools and formats accordingly.

▸ *Long-term investment*: Ironically, the derivative *may* be a long-term investment. Especially as formats that increase versatility and functionality develop, some choices may offer greater long-term functionality to the user. For text, this phenomenon has been borne out by XML, which libraries use for archival text and metadata: the emergence of popular browsers supporting XML display has made storage and retrieval of XML even more fruitful. Rich derivatives capable of supporting a wide array of applications will continue to emerge, making it possible for the *derivatives* to be long- or at least medium-term investments. However, derivatives have a significant interplay with the prime long-term investment, the master. A use format that makes it possible to derive surrogates in real time can privilege the master in the online system and allow a tighter coupling of master and system. This in turn contributes to better long-term support for the master.

▸ *Performance*: Access systems present dilemmas in balancing demands on the host system and on the user's desktop. Frequently, the choices for the host system and the user's desktop are intertwined. For example, image processing may be off-loaded from the server to the user's desktop. Or images may be processed on the server in order to diminish the need for specialized image tools or computing-intensive approaches on the user's desktop. In every case, however, decisions must take into account the speed with which the information can be delivered, the amount of computation required on the host system, and the speed with which users can view images.

Examples of Fitting the File to the Purpose

Panning and Zooming with Rich, Tonal Images   Several major digital library enterprises have adopted server-side wavelet compression for panning and zooming of high-resolution or large-format images.[9] At the Library of Congress, wavelet compression has been used effectively to display large-format images of materials such as maps.[10] David Allen explores many similar projects and their strategies for incorporating support for large-format materials.[11]

The University of Michigan's Image Services, part of the DLPS, uses server-side wavelet compression to allow users to pan and zoom in on a wide variety of high-resolution images, including art and architectural images.[12] From a single high-

9. Steven Puglia, "Fractal and Wavelet Compression," *RLG DigiNews* 2, no. 3 (June 15, 1998), www.rlg.org/preserv/diginews/diginews23.html.

10. Carl Fleischhauer, "Digital Formats for Content Reproductions" (July 13, 1998), memory.loc.gov/ammem/formats.html; Caroline Arms, "Historical Collections for the National Digital Library: Lessons and Challenges at the Library of Congress," *D-Lib Magazine* (May 1996), www.dlib.org/dlib/may96/loc/05c-arms.html.

11. David Yehling Allen, "Creating and Distributing High Resolution Cartographic Images," *RLG DigiNews* 2, no. 4 (August 15, 1998), www.rlg.org/preserv/diginews/diginews2-4.html.

12. Image Services uses LizardTech's MrSID™ Portable Image Format™ to support its work with wavelet compression: LizardTech home page, www.lizardtech.com.

resolution wavelet-compressed image, users can view dynamically generated JPEG derivatives at various resolutions, or can choose to navigate the highest-resolution image by viewing segments. The "master" wavelet-compressed image on the server is in the same resolution as the digital master (i.e., the fullest possible resolution), thus making possible a wider range of resolutions for derivatives, as well as extremely high-resolution zooming. By storing the master wavelet-compressed image on the server, Image Services diminishes the need to create new derivatives, and in fact can avoid this possibility until an entirely new online format is selected for its services. Nearly all processing is performed automatically by machine, and manual steps (e.g., executing programs for compressing images) can be performed by the organization's support staff. Image Services has avoided requiring users to acquire a specialized viewer for wavelet-compressed images and instead generates images in the more open JPEG format on the server at the time of the user's request. Although this requires more programming and computing resources on the server end, it reduces demand on the user's system and can make viewing an image much faster as well. This approach is not unique to wavelet compression and is also possible with formats such as FlashPix™, GridPix, and JTIP.[13]

Delivery of Images as Just-In-Time Derivatives Bitonal TIFF images of text pages in the ITU Group 4 format are ideal both as master files and as a part of an access strategy. Because of their extremely compact representation of information (e.g., typical images are rarely larger than 100 KB) and the widely available processing tools, several initiatives employ these bitonal TIFF images as an integral part of their online systems. This is by no means always the case, however. Important efforts such as Oxford University's Internet Library of Early Journals (ILEJ) and Pricing Electronic Access to Knowledge (PEAK) at the University of Michigan are relying fruitfully on the use of derived files.[14] Nevertheless, the opportunities presented by this rich, malleable format and compression will probably continue to spur the development of systems that use this archival master to derive a variety of other formats based on user demand.

One case where the ITU Group 4 bitonal TIFF is used in the online system to create multiple derivative formats is the University of Michigan's MOA system, subsequently adopted at Cornell. Key to the strategy employed by the DLPS is Doug Orr's image conversion tool, tif2gif.[15] Tif2gif generates GIF images at a variety of resolutions directly from the TIFF master files. By optimizing its performance around typical print-oriented bitonal information, Orr was able to generate GIF images from TIFF masters on a relatively low-powered computer in a matter of a few seconds. This performance is combined with intelligent strategies in the MOA system; for example, when an image is requested, a "look-ahead" (i.e., caching) converts the following page on the assumption that it is more likely to be requested. Other formats are offered, among these the real-time derivation of PDF files for printing. When a user prints an image, a locally developed tool processes the TIFF image to create a PostScript® file. This in turn is sent to Adobe Distiller®, which presents the PDF file to the user, also in a matter of seconds. As many as six different output options are available, all but two derived from the TIFF image on user demand. The others—a thumbnail format and OCR (optical character recognition) output—are derived offline.

The strategy taken in Michigan's MOA system is extremely efficient, but—as currently implemented—it is not without its deficiencies. While the system handles more than 100,000 image requests per month, it is possible that it could be overloaded by high demand.[16] Tif2gif introduces another more

13. Martin Rigby, "What is Flashpix™?" *FlashPix.com*, www.flashpix.com. For information about Gridpix, see "The Zoom Project," now.cs.berkeley.edu/Td/zoom.html. "JTIP (Images Pyramidales en Tuiles JPEG)," www.argyro.net/demo/avelem/jtip.htm.

14. "Internet Library of Early Journals (January 1996–August 1998) A Project in the eLib Programme: Final Report" (March 1999), www.bodley.ox.ac.uk/ilej/papers/fr1999/fr1999.htm. In the PEAK project, the DLPS compresses TIFF G4 images using the proprietary Cartesian Perceptual Compression to conserve disk space: "Cartesian Perceptual Compression: Technical Information," *Cartesian Products, Inc.*, www.cartesianinc.com/Tech. The decision to use CPC compression came as disk consumption was running in the hundreds of gigabytes per year, and because the University of Michigan is not expected to own and store the master versions of these images.

15. Doug Orr developed tif2gif for Michigan's implementation of the Elsevier TULIP journals. Source code and binaries are available online: "Tif2gif," kalex.engin.umich.edu/tif2gif.

16. The JSTOR project grappled with cascading performance issues when it began offering PDF as a print output for that system.

intriguing deficiency with its addition of grayscale to bitonal images. At a low resolution, this capability makes the image much easier to read than at, for example, a corresponding resolution in a PDF file. It can be a problem, however, when the image contains nontext information, such as illustrations. The addition of unnecessary or inappropriate grayscale makes halftones muddy and unattractive. This deficiency could easily be overcome by a conditional strategy to detect the presence of halftones and use another image conversion tool such as ImageMagick. Despite this drawback, using the master TIFF image as part of the access system has offered a great deal of versatility to users with no significant compromises.

The Case for Real-Time Creation of Derivatives

As collections of digital objects grow, the ability to scale the systems that provide access to them will depend on an efficient integration of the master file and the access version. This will sometimes result in employing the master file as part of an access strategy, deriving use versions upon user request, as described above. It may in fact result in employing the master file as the access file (i.e., using on-demand derivatives for display), as is the case in Michigan's Making of America. And in many cases, it will result in deploying use versions or other derivatives with a relatively long viability for user needs (i.e., not using the master as part of the access system). It could also result in the creation of an access master derived from the digital master that takes advantage of current capabilities for delivering images.

The argument for creating on-the-fly or real-time derivatives from a master file is driven by a rapidly changing environment. A fuller discussion of the rationale for this strategy can be found elsewhere.[17] At the heart of this strategy is the recognition that, as collections grow, patterns of use will parallel the patterns of use for print collections.[18]

That is, a small minority of the materials serves the vast majority of the uses. This pattern of use holds even for a popular, heavily used reference work, a large collection of journal articles, or a historical resource.[19] Data on use of large digital collections continue to support this thesis. At the same time, as networking, desktop, and display technologies change, users demand—and are better prepared to use—richer files in typically newer formats. Access systems are, quite simply, challenged by a rapidly changing environment.

A strategy that incorporates the digital master in the online system and generates just-in-time derivatives offers the digital library manager the opportunity to devote important system resources to caring for the digital master rather than the derivative and to serving user needs more flexibly. If indeed small proportions of the material are used, strategies can be aimed at serving actual use rather than statistically less likely potential use. This is not to say that the online systems will not be heavily used; in fact, use of nearly all digital collections is high, though only small proportions of the collection may be used. Real-time conversion can take advantage of the rapid shifts in technology. For example, in Michigan's MOA system, the lowest-resolution GIF image (i.e., less than 72 dpi) was a relatively popular choice at the outset of the public availability of the materials; within a year, use of the lowest resolution had fallen so much that it was dropped. In this case, system resources were not devoted to needlessly generating 650,000 derivatives, a tiny fraction of which would have been used. Between 1999 and 2000, the collection will grow to approximately three million pages, amplifying the importance of the decision several times.

The ability to adopt a "just-in-time" strategy depends on the nature of the source materials and

17. John-Price Wilkin, "Just-in-Time Conversion, Just-in-Case Collections: Effectively Leveraging Rich Document Formats for the WWW," *D-Lib Magazine* (May 1997), www.dlib.org/dlib/may97/michigan/05pricewilkin.html.

18. Allen Kent et al., *Use of Library Materials: The University of Pittsburgh Study* (New York: M. Dekker, 1979).

19. Price-Wilkin, "Just-in-Time Conversion." Preliminary research by PEAK found that 37% of roughly 1,100 journals got 80% of the usage ("Percentage of Collection Usage," www.lib.umich.edu/libhome/DLI/presentations/cni99/sld031.htm). However, relatively few articles were used from most of those journals. In fact, examining usage by article shows that an extraordinarily small percentage of the roughly 900,000 articles were used over the same period ("Repeated Accesses of Articles," www.lib.umich.edu/libhome/DLI/presentations/cni99/sld026.htm). Wendy P. Lougee, "PEAK: Pricing Electronic Access to Knowledge" (paper presented at Coalition for Networked Information, April 1999), www.lib.umich.edu/libhome/DLI/presentations/cni99/ sld001.htm.

Table 6.1. Conversion Times Per Image

	Average Time	Median Time	Average GIF	Median GIF	Average TIFF	Median TIFF
Tif2gif	3.98 sec.	3.9 sec.	78.1 KB	83.0 KB	84.1 KB	89.1 KB
ImageMagick	44.12 sec.	35.8 sec.	217.4 KB	228.9 KB		

Table 6.2. Comparative Times for Converting an Image Collection

	Images	Derivatives	Avg Minutes	Total Days	Total Years
Tif2gif	1,000,000	3	0.0663	138.22	0.38
ImageMagick	1,000,000	3	0.7353	1,531.98	4.20

the availability of tools for building the access system. The many available choices present challenges and opportunities. To demonstrate these challenges, as well as the perils of precomputing a large number of derivatives, data from Michigan's Making of America system were converted using two popular image conversion tools and the data were applied to a hypothetical one million-page collection. A sample of 53 600-dpi TIFF bitonal images was selected from MOA materials. The pages contained varying amounts of text, as well as blank pages, which commonly occur in these sorts of conversion projects. Table 6.1 provides the average and median times (in machine-computed seconds) for each of the conversions, as well as the ultimate size of the converted file.[20] Average and median sizes are provided for the TIFF images as well, for comparison.

Table 6.2 extrapolates the average time to convert these TIFF images, to calculate the amount of time needed to convert one million images to three derivative forms (e.g., high, medium, and low resolutions). For tif2gif, the time to create all three million images is approximately four and a half months. ImageMagick would take more than four years to perform the same operation. Converting Michigan's Making of America collection would take between one and twelve years, using this type of computer.

This comparison is not presented as an argument for tif2gif over ImageMagick. In fact, as discussed earlier, tif2gif is clearly not always applicable for converting bitonal TIFF images and ImageMagick has proven to be extremely versatile and reliable. Rather, with either tool, it takes a long time to convert a large collection and will take longer as collections grow. The likelihood that only a small proportion of the materials will be used in a given year raises questions about devoting this amount of system resources to the task, as well as questions about a strategy that would unnecessarily withhold these collections from users for such a long time. It is possible with today's technology to convert many digital images on demand, resulting in converting only those materials that are used. Moreover, by employing caching such as the look-ahead in MOA, the advantages of converting once for multiple uses can be achieved as well. When on-demand conversion is possible, its advantages are compelling.

Just-in-time strategies are not always possible with image collections. For example, 100-MB color TIFF images take considerable system resources and time to convert to derivative formats. The future for these sorts of images is currently being shaped. The strategies being taken with new formats and compression techniques reflect the hope of investing in future standards. The just-in-time strategy will take a new form, with the master file containing its own derivatives. Formats such as FlashPix, GridPix, or JTIP, or compression processes such as fractal, wavelet, and JPEG 2000, will enable access to both smaller and more detailed views from a single file. This file may be the digital master or an intermediate "access master" to store and manage for all types of images.

20. Conversions were performed on a Sun™ 143-Mhz Ultra-SPARC1 with a single processor and 256 MB of RAM under a low load.

BRINGING INTELLIGENCE TO IMAGES—ENHANCING ACCESS WITH TEXT

Putting Your Metadata to Work in the Online System

Basic metadata of all types—descriptive, structural, and even administrative—are pivotal in providing access to images. For a fuller discussion of the types and value of metadata, see chapter 5. Two points worth mentioning here are that enhancing access to image files depends significantly on the successful integration of metadata, and that metadata is not an incidental byproduct of capture—metadata does not just happen. Since even technologies such as OCR are a *type* of metadata, access to digital objects is, ultimately, a function of metadata.

Using Text to Enhance Access to Images

From humanities text-encoding initiatives to JSTOR, the strategy of combining images and text has been explored and debated. This debate, focusing on the relative value of encoded text or image (each typically to the exclusion of the other), has largely abated, not least because of the relatively low cost of capturing and storing images as part of an encoded-text strategy. For example, at the Library of Congress, an image and the encoded text of the page are viewed as complementary. And at the University of Michigan DLPS, low-cost imaging forms the backbone of all encoded-text conversion efforts. The benefits are two-fold: 1) users are able to consult images when the text resists straightforward encoding, and 2) online images provide a readily available source of text to staff and keyboarding firms for input and quality control of transcription and encoding. User feedback on the need for full text in access systems is overwhelmingly emphatic and enthusiastic.[21] Many strategies are available for adding full text to image systems, including uncorrected OCR, corrected OCR, and keyboarded text. Cost, usability, audience, and source materials are key factors in determining which of these approaches, if any, is

most appropriate. Ultimately, however, better text will result in better retrieval.

The range of factors to consider is intimately tied to the materials and their intended audiences. For example, a single-page manuscript may be so short and well known that searching is not an issue; what counts is an accurate representation of its physical appearance. When accommodating vision-impaired users, the need for highly accurate text will outweigh the value of a graphical representation of a page. Textual and linguistic analysis, historical study, and large-scale retrieval, among many areas of use, introduce factors that should be considered in creating encoded-text versions of print materials.

OCR, Corrected OCR, and Keyboarding: Costs

Although many elements contribute to the difficulty or ease of capturing text from print, a deciding factor may be the cost of fully accurate capture. One might well begin any discussion of accuracy in capture by asking, How *perfect* is *perfect*? In outsourcing work to service bureaus, accuracy is measured as the percentage of characters (including single spaces) accurately reflected in the resulting electronic text. The highest and most expensive level of accuracy from such firms is typically 99.995%, or no more than one error in 20,000 characters of electronic text (approximately 20 printed pages).[22] Anecdotally, many traditional editorial enterprises with extremely high standards are unable to exceed this level of accuracy.[23] Thus, this discussion uses the 99.995% figure as equivalent to perfect, or fully accurate, text.

Alternative methods of conversion can achieve results with different costs and accuracy, depending on the source text. The highest level of accuracy is frequently achieved cost-effectively only through keyboarding. Service bureaus employ a variety of methods, including independent key-

---

21. Bonn, "Building a Digital Library."

22. Service bureaus measure accuracy and charge by the byte rather than by the character. Charges for data entry include, for example, spaces, carriage returns, and markup.

23. For example, in converting the *Middle English Dictionary* (MED), a monument of 20th-century lexicography, the University of Michigan used a service bureau for keyboarding the printed volumes. Nearly all errors found in proofing the keyboarded text were traceable to the printed MED.

OCR Trends and Implications

Kenn Dahl
President and Founder, Prime Recognition

Optical character recognition (OCR) transforms "nonintelligent" raster images of text pages into "intelligent" electronic text that can be searched, displayed, manipulated, or transmitted much more efficiently. To date, OCR has technical limitations that have hampered its general applicability in document-conversion projects. Poor scanned image quality and complex layouts, which challenge OCR, unfortunately are quite common. Our calculations in 1998 indicated that less than 20% of all imaging projects used OCR. But the technology is evolving rapidly, with strong implications for document-conversion projects.

Character Accuracy

The primary measure of OCR performance—and its key limitation—is accuracy. Character accuracy, the most important aspect of text recognition, varies widely based on the quality of the image. On typical machine-produced documents, the leading OCR engines currently generate about 98–99.5% accuracy, which translates to 5 to 20 errors per 2,000 characters. However, character-recognition accuracy is improving rapidly:

- Core recognition technologies are improving by approximately 10–15% per year.

- Among new high-end technologies is a technique called "voting," which combines multiple recognition technologies for higher accuracy. This technology generates 65–80% fewer errors than the best conventional stand-alone OCR engines.

- Improved technologies that create a cleaner image prior to OCR include better scanners and more automatic and intelligent image-enhancement software.

- Technologies that clean up OCR errors have also improved, including automated error-reduction technologies such as lexical checking and more efficient manual tools.

OCR error rates are 50% lower than several years ago and this trend will continue. For example, three years ago the best conventional OCR products generated about 20 errors on a typical 2,000-character page. Today, the best conventional OCR products generate ten errors, and in three years they will generate about five errors on the same page.

Format Accuracy

Replicating the exact format of the original image is very difficult for OCR. Significant progress has been made in this area in the last three to four years, particularly for the Rich Text, MicroSoft® Word, and PDF Normal formats. However, many users still do not feel comfortable with the quality of the OCR format and have moved towards hybrid formats that combine the original image with searchable text, such as PDF. This hybrid format has the advantage of faithfully replicating the look of the original document, which often has useful legal implications, as well as reducing the requirement for exceedingly high OCR accuracy rates. A key disadvantage to hybrid formats has been their file size, due to the inclusion of the raster image.

New Image Types

Historically, OCR was designed for business documents with fairly large, readable fonts. Hence, OCR was optimized for 300-dpi images. Over the last two to three years, however, support for 400- and 600-dpi resolution has improved substantially, which is critical for documents with small font sizes. A 600-dpi scanning resolution, for example, usually improves OCR accuracy by 30–50% over 300-dpi scanning for text printed in 7-point font.

Business documents are fairly simple black-and-white pages, so OCR has historically been focused on 1-bit images. OCR companies have experimented with color and grayscale image support, but with poor results until very recently. Only in 1999 did color and grayscale support improve to the point that the technology could be considered practical. Color images, for example, can consume 50 times more RAM, can double or triple OCR times, and can hurt OCR accuracy as much as help it. However, the availability of inexpensive RAM, faster CPUs, and improved OCR technologies to process color data means that OCR can now finally be used on color images.

Declining Hardware Costs

OCR is very computing intensive. The cost of PC-based MIPS (millions of instructions per second) has been decreasing at about 30% per year. What used to require a $50,000 custom hardware box can now be achieved on a $2,000 PC with an Intel® Pentium® III processor. In most production environments, the cost of PC hardware and OCR software is minor compared to the labor costs of project design/integration, ongoing management, and any manual processing steps, such as scanner loading and OCR cleanup.

Production OCR Engines

For more than a decade, OCR product development has been driven by the large desktop/single-user market. Recently, a new submarket has emerged to serve high-volume users. Developments that serve this market include:

- Dedicated OCR servers (scanning, verification, and other tasks are performed elsewhere)

- Unattended batch processing

- More reliable software

- Facilitation of scalable architecture (e.g., multiple CPUs on the same machine or multiple CPUs on a network)

- Higher accuracy through a choice of OCR engines (matching the best engine to the page type), or voting

CONTINUED ON PAGE 112

boarding by two or more individuals and subsequent machine-based comparison (referred to as double keying, triple keying, etc.). Ultimately, the methods matter far less than the resulting product. A large and growing industry performs this work. A useful annual review of service bureaus is published in *Imaging & Document Solutions* each March.[24] Alternatives to keyboarding include uncorrected OCR, OCR corrected through a single pass of software-mediated proofreading, and fully corrected OCR.

Achieving Highly Accurate OCR

OCR can frequently capture text from images effectively and inexpensively. Uncorrected OCR cannot approach the accuracy of keyboarding and in many cases correcting OCR to reach 99.995%

24. *Imaging & Document Solutions*, www.imagingmagazine. com.

accuracy is more expensive than keyboarding. Cost and accuracy depend on staffing and software, but are overwhelmingly based on the quality of the source text.

Uncorrected OCR, which has been employed by a number of major projects such as JSTOR and Michigan's Making of America, is the least expensive way to add text to aid in retrieval. Its accuracy may approach that of keyboarding, but varies widely depending on factors such as the quality and size of the fonts in the source text. Table 6.3 demonstrates some of this variation for five sample pages with different textual characteristics that were processed with an OCR package, without operator intervention. They include:

▸ *Complex page*: The image contains a table of contents with 6- to 10-point type and complex layouts. Variation in point sizes on a single page typically results in lower levels of accuracy. As most tables of contents share the characteristics

CONTINUED FROM PAGE 111

Trend toward Outsourcing

Because OCR technology is changing rapidly, it can be difficult to stay current with the latest advances for increasing accuracy and/or reducing costs. The increasing availability of high-bandwidth communications channels favors the development of OCR services over products, eliminating the need for the user to become a state-of-the-art expert. Today, third-party service bureaus perform much OCR work. Typically, the service bureau accepts images via CD or tape, resulting in significant delays for shipping data. These delays limit outsourcing for time-sensitive projects. Some service bureaus have begun to exchange data via FTP, but projects tend to be fairly small since many users do not have very high speed network connections. Very large-capacity, state-of-the-art OCR services that are readily available to the user through an Internet connection will be popular in the future. Note, however, several hurdles to their acceptance: 1) scanning and other processes dealing with the physical page may still be done locally, and 2) as fast as Internet access speeds are growing, demands on that bandwidth will also increase for larger projects and larger files.

Support for Broader Character Sets

Historically, OCR has focused on the English-language character set. As the market has expanded, a broader set of languages has gained more attention, although the vast majority of products still support only Roman-language character sets. Although most products now support "training," or the ability to learn a new font, the training can help the product only to a limited extent. Therefore, in practice, despite these advances OCR still does not perform well with more stylized type styles, such as gothic, that are significantly different from modern ones.

Viability of Handprint?

First the good news: Much progress has been made in improving recognition rates of handprint and, to a lesser extent, handwriting. The bad news is that accuracy (usually 85–95%) is still more than an order of magnitude worse than machine-print recognition. The only applications that lend themselves to handprint recognition currently are those in which the data are highly constrained; for example, medical insurance forms or financial checks. Recognition of free-form (unconstrained) handprinted or handwritten text is still decades away.

OCR Resources on the Web for Further Research

▸ "Character Recognition (OCR/ICR)," *Yahoo*, dir.yahoo. com/Business_and_Economy/Companies/Computers/ Software/Character_Recognition__OCR_ICR_/. A fairly complete list of OCR engine vendors and resellers.

▸ "OCR and Text Recognition: Commercial Research and Products," hera.itc.it:3003/~messelod/OCR/Products. html. Current and complete list of OCR engine vendors, plus links to other OCR resources.

▸ "What Is the Best OCR? OCR Software/Scanner Survey Results," www.hosc.net/ocr. Well-organized lists of desktop and production OCR products, along with OCR-related coverage of scanners, image enhancement software, etc.

▸ comp.ai.doc-analysis.com. Newsgroup for OCR questions and answers; however, tends to be dominated by new users asking questions about desktop issues.

Table 6.3. OCR Accuracy Results[25]

	Complex Page		Normal Page		Poor Page		Small Facsimile		Small Note	
DPI	300	600	300	600	300	600	300	600	300	600
Accuracy (%)	98.36	99.52	99.92	99.94	99.24	98.82	0.00[26]	94.23	N/A[27]	99.13
Characters	5,420		5,038		6,949		4,485		1,719	
Errors										
Total	89	26	4	3	53	82	4485	259		15
Char.	76	11	0	0	53	82		250		9
Space	3	3	4	3	0	0		8		2
Num.	10	12	0	0	0	0		1		4
Confidence	805	831	897	897	875	872	266	740	701	849

of the sample page, accuracy rates on tables of contents are typically low. (The need to reliably retrieve content on a table of contents page via OCR may not be great, however, since the pages themselves will typically be retrievable by structural metadata.)

‣ *Normal page*: The image is a typical example of typefaces used in periodicals during the latter half of the 19th century. This page, from an 1897 issue of *The Century Magazine*, is printed in 10-point type and consists of a simple two-column layout with no illustrations, titles, or footnotes. The type is printed clearly and exhibits consistent ink density across the entire page. This is the best type of material for OCR programs.

‣ *Poor page*: Although the type size and simplicity of layout on this page are consistent with the sample of the normal text page, the printing is

poor. Nonetheless, the OCR program performed remarkably well on this sample page.

‣ *Small facsimile*: This is a copy of an 1849 periodical page, reduced significantly to appear on the page of this later 19th-century journal, resulting in approximately 3-point type. The reproduction and reduction contribute to the poor quality of the OCR at 600 dpi, and proved impossible for the 300-dpi sample.

‣ *Small note*: This note for the facsimile above has characteristics of typeface and size equivalent to 6-point Times Roman. Although the OCR program could read the 300-dpi sample, the accuracy was extremely poor (approximately 50%).

The results of this test point to several conclusions.

‣ Higher-resolution imaging (i.e., 600 dpi rather than 300 dpi) nearly always leads to better OCR results. Increasingly, OCR programs can effectively support 600-dpi image files, broadening the range of fonts that can be processed.

‣ Mixed typefaces and complex layouts may lead to noticeably lower results.

‣ "Garbage in, garbage out": poor image capture leads to poor results. Radically skewed pages or

25. The results in table 6.3 were generated using PrimeRecognition OCR software on a variety of test pages at two resolutions (300 dpi and 600 dpi). The OCR results were not corrected, and no manual mediation (e.g., OCR "training" to recognize types of characters) improved accuracy. The elements in the table include: 1. DPI: Optical resolution at which the pages were captured. 2. Accuracy (%): The rate of accuracy obtained by dividing the total number of errors by the total number of characters found on the page, expressed as a percentage. 3. Errors: Total errors—the sum of character errors (missing or mistaken alphabetical characters), space errors (where spaces were introduced in the middle of words or where a space between words was omitted), and number errors (missing or mistaken numerical characters). 4. Confidence: PrimeRecognition software generates a value (1–900) for its confidence in recognition. There is generally a strong correlation between confidence and accuracy. This figure is provided only for comparison in the table.

26. The OCR package was unable to recognize any characters in the 300-dpi sample. Its documentation noted this limitation.

27. The accuracy rate of the 300-dpi image for this page was approximately 50%; because of the high number of errors or possible errors, a precise measure was difficult.

blurred text due to limitations in scanning will lower the OCR accuracy rates considerably.

‣ Smaller typefaces and second-generation impressions have a detrimental effect on the quality of capture.

‣ Extremely reliable capture (e.g., 99.94% accuracy) is possible for "normal" late 19th-century text written in English.

Improving Accuracy: Two Heads are Better Than One

OCR software's ability to recognize a character depends in part on the strategies taken by the software designers.[28] Algorithms for pattern recognition vary from manufacturer to manufacturer, and

28. Peter Wayner provides an extremely useful overview of the technology and different industry approaches in Peter Wayner,

while a package may excel in one area, it may prove much less effective in another. For example, the strategy taken by ExperVision's TypeReader® is perhaps ideally suited for older printed texts with, for example, broken type.[29] Two text-conversion projects that included large amounts of 19th-century material independently concluded that TypeReader offered the most effective retrieval for their materials.[30] Researchers at the University of Nevada, Las

"Optimal Character Recognition," *Byte* 18 (December 1993): 203–4.

29. Wayner, "Optimal Character Recognition."

30. From 1994 to 1995, the American Verse Project evaluated six different OCR packages before choosing TypeReader® for the bulk of its work. At the same time, Digital Imaging, JSTOR's service bureau, selected TypeReader to convert the significant body of older journal material. Both JSTOR and the American Verse Project have since moved to other methods for character capture.

Content-Based Image Retrieval

John P. Eakins
Director, Institute for Image Data Research, University of Northumbria at Newcastle

Traditional means of classifying and indexing image collections include general-purpose schemes such as MARC, special-purpose schemes such as ICONCLASS, and in-house, keyword-based systems.[1] Although in the right hands such methods can provide remarkably effective retrieval, they suffer from a number of drawbacks. The most fundamental, as Howard Besser observes, is that, unlike books, images make no attempt to tell us what they are about and may often be used for purposes their originators did not anticipate.[2] Inevitably then, the terms used by human indexers often poorly match those used by searchers.[3] Indexing is also time consuming: 7 to 40 minutes for a still image according to the literature. Such problems have stimulated interest in automatic methods for indexing and retrieving images—a field now generally known as content-based image retrieval (CBIR).[4]

How Does CBIR Work?

Virtually all current CBIR systems operate in a similar fashion. Each image entered in a CBIR database is analyzed automatically to extract features that characterize its appearance. The features most commonly chosen are numerical measures of shape, color, and texture, though more complex mathematical transformations of pixel values have also been used successfully.[5] The set of feature values characterizing each image—often known as a feature vector—is then stored in the database.

The most common type of search supported by CBIR systems is query by example: users submit an example of a desired image and the system attempts to find similar images by comparing feature vectors of stored and query images.

How Effective is CBIR?

CBIR techniques are currently most effective for retrieval by image *appearance* (such as finding fabrics of a particular color in an online mail-order catalog), rather than image *semantics* (such as identifying a picture of a child playing with a dog). Some semantic queries can be formulated in terms of color, texture, or shape, given a little ingenuity on the part of the searcher. For example, photographs of beach scenes might be retrieved by specifying images that are 75% yellow and 25% blue, and pictures of elephants by drawing a sketch of an elephant on the screen. But images are still matched purely by superficial appearance, with no understanding of image content. Hence, our elephant query could well retrieve pictures of, say, insects shaped like the query sketch, while missing pictures of elephants in different poses. True semantic image retrieval remains an elusive goal, though it is actively being investigated.[6]

Available CBIR Software

The most well-known, general-purpose commercial CBIR systems are:

‣ *QBIC*® (Query by Image Content) from IBM: This offers retrieval by any combination of color, texture, shape, or text keyword. Image queries can be formulated by selecting from a color palette, specifying an example query image, or sketching a desired shape on the screen. For an online demonstration of the software, see wwwqbic.almaden.ibm.com.

CONTINUED ON PAGE 115

CONTINUED FROM PAGE 114

▸ *The VIR Image Engine™* from Virage Inc.: This is available as a series of independent modules that systems developers can build into their own programs, making the system easy to customize for specialist applications. Further information about the VIR Image Engine can be found at www.virage.com/products/image_vir.html. Virage also provides the technology for the AltaVista™ Photo & Media Finder (image.altavista.com/cgi-bin/avncgi).[7]

▸ *Visual RetrievalWare®* from Excalibur Technologies: This offers a variety of image indexing and matching techniques based on the company's own proprietary pattern-recognition technology. For more information on Visual Retrieval-Ware, see www.excalib.com, and a demonstration of the Yahoo!™ Image Surfer, which uses Excalibur technology, at isurf.yahoo.com.[8]

In addition, demonstrations of several experimental CBIR systems can be found on the Web.[9]

Applications of CBIR

Potentially, CBIR techniques have a wide range of applications, including law enforcement, intellectual property, architectural and engineering design, fashion and interior design, journalism and advertising, medical diagnosis, education and training, and cultural heritage. Users in many of these areas are actively evaluating CBIR techniques to see how useful they are likely to prove in practice. Very few have yet put CBIR into routine use.

Not surprisingly, CBIR has made the most impact in areas where retrieval by appearance, rather than semantics, is of prime importance, such as fingerprint matching and trademark image registration. CBIR has made less impact in applications where the *meaning* of an image is important, such as retrieval of images from photo libraries to illustrate newspaper or magazine articles, or identification of works of art dealing with a particular theme. IBM's QBIC has been used experimentally in a number of California art libraries.[10]

The Way Ahead

CBIR is an exciting but immature technology that is still primarily the domain of the research scientist. Few operational image archives have yet shown any serious interest in adopting it, mainly because of the fundamental mismatch between the needs of most users and the capabilities of the technology.

Given the amount of research currently devoted to developing better CBIR techniques at all levels, this situation may not last for long. The latest generation of experimental CBIR systems already uses techniques such as relevance feedback to model users' needs more accurately, and within a few years current work on object recognition could well bridge what is sometimes known as the semantic gap.[11] CBIR is likely to become established in niche applications quite soon. Its more general adoption will have to await significant advances in technology—but these could happen sooner than many people think.

Notes

1. C. Gordon, "An Introduction to ICONCLASS," in *Terminology for Museums: Proceedings of an International Conference Held in Cambridge, England, 21–24 September 1988*, ed. D. Andrew Roberts (Cambridge, UK: Museum Documentation Association with the assistance of the Getty Grant Program, 1990), pp. 233–244.

2. Howard Besser, "Visual Access to Visual Images: The UC Berkeley Image Database Project," *Library Trends* 38, no. 4 (1990): pp. 787–798.

3. M. Flickner et al., "Query by Image and Video Content: The QBIC System," *IEEE Computer* 28, no. 9 (1995): pp. 23–32; R. Manmatha and S. Ravela, "A Syntactic Characterization Of Appearance and Its Application to Image Retrieval," in *Human Vision and Electronic Imaging II*, Proceedings of SPIE—The International Society for Optical Engineering 3016, ed. B. E. Rogowitz and T. N. Pappas (Bellingham, WA: SPIE, 1997): pp. 484–495.

4. V. N. Gudivada and V. V. Raghavan, "Content-Based Image Retrieval Systems," *IEEE Computer* 28, no. 9 (1995): pp. 18–22.

5. M. Flickner et al., "Query by Image and Video Content: The QBIC System," *IEEE Computer* 28, no. 9 (1995): pp. 23–32; R. Manmatha and S. Ravela, "A Syntactic Characterization Of Appearance and Its Application to Image Retrieval," in *Human Vision and Electronic Imaging II*, Proceedings of SPIE—The International Society for Optical Engineering 3016, ed. B. E. Rogowitz and T. N. Pappas (Bellingham, WA: SPIE, 1997), pp. 484–495.

6. D. A. Forsyth et al., "Finding Pictures of Objects in Large Collections of Images," in *Digital Image Access and Retrieval: 1996 Clinic on Library Applications of Data Processing*, ed. P. B. Heidorn and Beth Sandore (Urbana, IL: Graduate School of Library and Information Science, University of Illinois at Urbana-Champaign, 1997), pp. 118–139.

7. A. Gupta et al., "The Virage Image Search Engine: An Open Framework for Image Management," in *Storage and Retrieval for Image and Video Databases IV*, SPIE Proceedings 2670 (Bellingham, WA: SPIE, 1996), pp. 76–87.

8. J. Feder, "Towards Image Content-Based Retrieval for the World-Wide Web," *Advanced Imaging* 11, no. 1 (1996): pp. 26–29.

9. John P. Eakins and Margaret E. Graham, "Content-Based Image Retrieval: A Report to the JISC Technology Applications Programme," (January 1999), www.unn.ac.uk/iidr/report.html.

10. Bonnie Holt and Laura Hartwick, "Retrieving Art Images by Image Content: The UC Davis QBIC Project," in *Electronic Library and Visual Information Research, ELVIRA 1: Proceedings of the First ELVIRA Conference, Held May 1994 at De Montfort University, Milton Keynes*, ed. Mel Collier and Kathryn Arnold (London: Aslib, 1995), pp. 93–100.

11. Y. Rui et al., "Relevance Feedback Techniques in Interactive Content-Based Image Retrieval," in *Storage and Retrieval for Image and Video Databases VI*, SPIE Proceedings 3312, ed. Ishwar K. Sethi and Ramesh C. Jain (Bellingham, WA: SPIE, 1997), pp. 25–36.

Vegas, reviewed OCR accuracy annually for several years. Most years, changes in the technology produced another front-runner, though different front-runners for each type of text tested. Nevertheless, they concluded that "voting" systems, which combine recognition technologies, always provided the highest accuracy.[31] A number of highly accurate voting systems are now on the market.

Improving Accuracy: Correcting OCR

OCR results can be improved through manual correction, regardless of the accuracy obtained through an initial pass of OCR. Most OCR programs are supplied with a proofing interface. The program assigns a confidence level to the value it gives each captured character. To proofread, the OCR operator can set a threshold for suspect characters based on the confidence level and then press the Tab key to move from one low-confidence character to the next. If the operator makes no correction, the reported value for the character is retained. A number of service bureaus routinely perform this single-pass OCR proofing and it may be an extremely cost-effective way to improve accuracy. The cost of single-pass proofing is simply a matter of time and labor, and will depend largely on the original accuracy of the automated OCR: the more errors, the higher the cost to correct the text. Combined with voting software, this process can produce highly accurate text at relatively low cost.[32]

Choices of methods must ultimately take into account desired outcomes, including costs. The power of the technology has transformed OCR as a tool in recent years, bringing high-quality text to images at relatively low cost. For less than the cost of imaging, Michigan's Making of America project typically achieves 99.8% accuracy for its texts with no operator intervention.[33] However, users and usability must be the first and final considerations and for a number of purposes, including textual

analysis and serving vision-impaired users, the value of fully accurate text may force the use of more expensive options.

Future Developments Bringing Intelligence to Images

As we strive to enhance access systems for our users, the technologies that are most successful will be those that bring the greatest intelligence and flexibility to the images we put online. As Yecheng Wu suggests in his sidebar, raster-to-vector conversion holds promise for making graphic information more easily manipulated. Ironically, this technology may also offer important functionality for textual material, especially for supporting graphics such as graphs and charts, as well as for retaining page layout in rich, generic ways.[34] As John Eakins argues in his sidebar, content-based image retrieval (CBIR) may provide important methods for image discovery, augmenting textual description or perhaps even reducing the need for it in the future. CBIR may provide similar advantages for text-based materials, with the key to finding illustrations or particular types of tables in online books and journals.

The ability to provide better access to textual materials will depend in large part on the availability of improved OCR techniques and different forms of OCR output. As outlined in Kenn Dahl's sidebar, some suggest that improvements in single OCR engines are unlikely to be substantial. Instead, voting techniques may someday make their way into popular software packages. More progress in improving access by the addition of text may come simply from incremental advances in already existing technology. For example, Adobe Acrobat Capture format, which merges recognizable text, image, and layout information in a single PDF file, is unlikely to provide better OCR for retrieval, but may be an excellent on-the-fly derivative to improve access for users. On-the-fly OCR itself has become feasible with higher processor

31. Stephen V. Rice, Junichi Kanai, and Thomas A. Nartker, "The Third Annual Test of OCR Accuracy," *ISRI Technical Report* no. 94-03, April 1994 (presented at SDAIR '94, the Third Annual Symposium on Document Analysis and Information Retrieval), pp. 11–28.

32. Initial unpublished data from the Humanities Text Initiative at the University of Michigan indicates that 99.995% accurate data can be achieved in more than 50% of texts by a single pass of OCR. DLPS has found that the cost for single-pass proofing an average page of MOA text is $0.25.

33. Douglas Bicknese, "Measuring the Accuracy of the OCR in the Making of America" (a report prepared in fulfillment of

directed field experience requirements, University of Michigan, School of Information, winter 1998), www.umdl.umich.edu/moa/moaocr.html.

34. Arthur Gingrande, "The State of Optical Character Recognition," *Imaging & Document Solutions* (December 1998), www.imagingmagazine.com/db_area/archs/1998/December/The_4_1696.htm.

Raster, Vector, and Automated Raster-to-Vector Conversion
Yecheng Wu
President/CEO, Able Software Corp.

Differences Between Raster and Vector Images

Raster and vector are the two basic data structures for storing and manipulating images and graphics data. All of the major geographic information systems (GIS) and computer-aided design (CAD) software packages available today are based primarily on one of the two structures, while they have some extended functions to support other data structures.

Raster images come in the form of individual pixels, each indicating a distinct spatial location and attribute, such as color. The resolution of the acquisition device and the quality of the original data source determine the spatial resolution of a raster image. Because all spatial locations must be represented by individual pixels, increasing resolution increases file size geometrically.

Vector data come in the form of points and lines that are geometrically and mathematically associated. Points are stored using the coordinates; for example, a two-dimensional point is stored as (x, y). Lines are stored as a series of point pairs, where each pair represents a straight-line segment; for example, $(x1, y1)$ and $(x2, y2)$ indicate a line from $(x1, y1)$ to $(x2, y2)$.

In general, the vector structure produces a smaller file size than a raster image because only point coordinates are stored in vector representation. The difference in file size becomes most dramatic in the case of vector images containing large homogeneous regions, where boundaries and shapes are the primary interest. Because vector images contain fewer data elements than raster images, they are easier to manipulate and scale. This makes the vector data structure the apparent choice for most mapping, GIS, and CAD software packages.

Another advantage vector images have over raster images is the flexibility of resizing without loss of resolution. For example, graphical features such as rivers and roads in a map viewed with a real-world projection system can be easily displayed at any scale without physically changing the data. By contrast, a raster image has to be stretched and distorted when scaled above its native resolution.

There are a few drawbacks to converting raster images to vector images. The vector files tend to be tied to specific program applications; they are not as effective at representing photorealistic documents; and they may require long redraw times, especially for complex images. Future development efforts are bound to address these issues, broadening the scope and use of vector images.

Challenges in Raster-to-Vector Conversion

An image source must be in vector format before it can be accepted by most GIS or CAD systems for processing and visualization. Although the vector data structure provides a simpler and more abstract data representation than the raster image, automatic conversion from raster to vector (vectorization) is not easy. In the past, vector data had to be generated by manual tracing or drawing on a digitizing tablet or onscreen. Today, automatic conversion of raster images to vector format has become the primary method for acquiring vector data, especially for large projects such as national-level mapping and GIS data-capture applications. A complete raster-to-vector conversion process includes image acquisition, preprocessing, line tracing, text extraction (OCR), shape recognition, topology creation, and attribute assignment.

The image-acquisition process generates the initial raster image at a certain spatial resolution. The quality of the original source and the raster image created from it are key factors in determining the quality and accuracy of the vectorized data. Therefore, it is critical that original source documents be scanned at a sufficiently high resolution to ensure adequate capture of all significant detail. Too low a resolution may merge features such as lines and curves, making it impossible to vectorize correctly. The impact of too low an image resolution will be most evident when vector images are scaled. If the scanning resolution is set too high, however, the resulting image file not only requires an unnecessary amount of system resources for processing, but noise and artifacts may be introduced, which will affect the quality of the vector image.

Most good-quality black-and-white maps and engineering drawings, including color map separates, can be scanned as 1-bit monochrome. Maps in poor condition, with staining or surface dirt, should be scanned as 8-bit grayscale and enhanced using imaging software to remove background and noise before vectorization.

Color documents, such as satellite and aerial photos, can be used to create vector data. High-resolution color scanning, especially for large-format originals, is still quite expensive, and the large file sizes require more system resources for storage and processing. In addition, color classification and color separation in vectorization are very sensitive to the color quality of the scanned image. There are many different types of color classification methods, but they are similar in principle, creating statistics from the color elements (red, green, and blue) of pixels and grouping the pixels into classes based on their color characteristics. Once a color image is classified, the colors can be separated into much fewer layers and vectorized layer by layer. For some color images, however, the colors from different features may be mixed up to create ambiguous color boundaries, resulting in incorrect classification and degraded accuracy in the final vector data.

Choosing the Right Conversion Tool

Recent developments in automated raster-to-vector conversion technology make it possible to scan a document and convert it into vector format in a matter of minutes or even seconds. This process can take days or weeks to complete manually, and the quality of the resulting vector file is limited by the skill of the tracer and the spatial resolution human hands can reach. Scientific tests have shown that the best resolution

CONTINUED ON PAGE 118

CONTINUED FROM PAGE 117

achieved by manual tracing is 40 dpi while an optical scanner can easily reach 1,200 dpi.[1]

Several raster-to-vector conversion software packages are available for various types of applications, such as engineering drawing conversion, map digitizing, and GIS data capture. For example, Able Software Corp. (www.ablesw.com) developed their R2V™ software in 1993 with a focus on vectorizing scanned maps and creating GIS data.

To select the right tool for the task, ask these questions:

▸ *Does it support different image types, such as bitonal, grayscale, and color?* This is quite important for color source images. Treating color images as black-and-white or grayscale loses all color information and significant editing may be needed to separate colors by hand.

▸ *Is it designed for maps or engineering drawings?* In practice, the handling of map data and engineering data is quite different although they both are vector-based. If a package is designed for CAD drawings, the algorithms normally work well for straight lines and regular geometric shapes, but are not efficient for curving lines, polygons, and topology between polygons.

▸ *Does it support the native format for your application?* It is unfortunate that most vector file formats used today are constructed differently and data exchange between two vector formats may result in some data loss. One format may be excellent for CAD data transfer, but very limited if you need to get data into a GIS or mapping database. When creating vector data, it is always better to use the native format that the target system supports.

Note

1. L. R. Poos and Yecheng Wu, "Digitizing History: GIS Aids Historical Research," *GIS World* (July 1995): pp. 48–51.

speeds and lower memory costs, but it is likely to have little value beyond presenting another format to users. Similarly, color and grayscale information may be used to generate OCR, ensuring the ready capture of textual information from images other than Group 4 TIFF images.[35] The future of improving access to images lies in two areas: increased accuracy for discovery, and improved flexibility of the resulting resource, empowering users with a more vital text.

Libraries have an extraordinary opportunity to enhance access to their collections through the processing and deployment of images. To do this effectively, we must mount an aggressive two-pronged strategy that privileges both the needs of our materials and the needs of our users. System strategies that account for long-term needs while creating derivatives for current users contribute significantly to the long-term cost-effectiveness of our access systems. The rich master image file and in particular building systems around that master are critical to that end. User studies are important before, during, and after the creation of access systems for images. Endowing our systems with flexibility and intelligence will ensure that our users are the greatest advocates of our access systems.

35. OCR developers occasionally announce support for color or grayscale documents (cf. Paravision, 1999) when in fact they are turning color information to binary information ("binarizing" it) and then performing OCR.

Plate 1. Color Shift Caused by Photography. *Left*, image scanned directly from the document; *right*, image scanned from positive film that has a red cast.

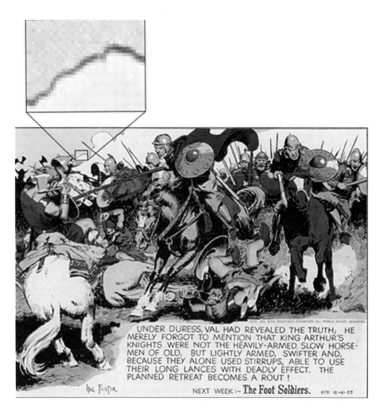

Plate 2. Adequately Rendered Stroke (color version of Fig. 3.7)

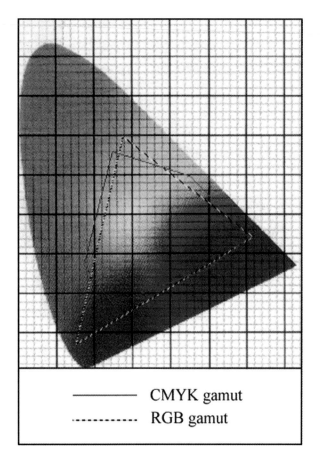

———— CMYK gamut

············ RGB gamut

Plate 3. Color Gamuts

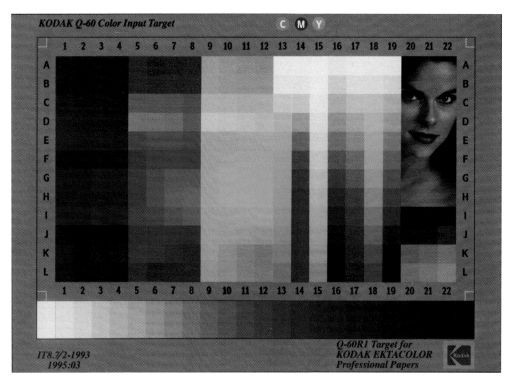

Plate 4. Kodak Q-60 Color Input Target (IT8). Courtesy of Eastman Kodak Company. Kodak is a trademark.

Plate 5. Measuring RGB Values. Color values can be measured with Adobe Photoshop software. After opening the Info palette (Windows/Show Info), position the pointer over any part of your image and the color value under the pointer will be displayed at the top left column of the palette. When you are viewing the reference bar, your image should be displayed in RGB mode (make sure that the left column reads RGB, not CYMK). Refer to the Adobe Photoshop manual to set the pixel area to read and interpret color values. See the user guide of the grayscale target to find the ideal density values for each color patch. This photograph was scanned with a Kodak Color Separation Guide (left) and Gray Scale (right). When the pointer is placed over the first black patch of the reference bar, the values read R: 5, G: 5, and B: 5. Ideally, this patch should be between 0 and 10. The middle dialog box illustrates how to read a neutral (or midlevel gray) patch (R: 96, G: 96, B: 96). Ideally, this patch should be 96, indicating pure gray. The last patch, which is the lightest (R: 255, G: 255, B: 255), should be between 240 and 255. Although the RGB values in this example are very close to their ideal values and consistently equal among the three channels, this is not always possible in the real world.

Plate 6. Color Shift. Note the effect of higher blue channel value (B) on the overall color appearance of the bottom image.

7 Image Management Systems and Web Delivery

Peter B. Hirtle
Co-Director, Cornell Institute for Digital Collections

WHAT IS AN IMAGE MANAGEMENT SYSTEM?

Digital imaging projects typically involve the creation of thousands, and sometimes millions, of digital files. Digital conversion itself, for example, usually produces at least one bit-mapped image file. Multiple smaller image files, derived from the original scan, may also be produced, as well as metadata about each image.

To organize, manage, and provide access to master images, derivative images, and the accompanying metadata, you need an image management system. Image management systems are usually built using databases and are frequently referred to as image database software. A database is not required for an image management system, however; other software tools can meet some of the same functional needs.

A successful image management system addresses multiple functions. Some of the functionality of an image management system is inherent in the software itself. Web browsers, for example, can identify and display common image file formats; other software may allow users to zoom in on images. Much of the functionality in an image management system, however, depends on the metadata that accompanies the image. In short, the image management system will be only as good as the metadata stored in it.

An image management system can:

- Record the storage location of master images and derivatives, usually by maintaining pointers to the file locations of images in the image management system. In some cases, the image itself is included in the database as a binary large object (BLOB).

- Search for and retrieve images.

- Provide an access interface, frequently via the Web.

- Provide a contextual framework for the digital images, such as an associated finding aid.

- Provide a structural framework for the digital images, such as table of contents.

- Track and control source material and generated images during the digitization process.

- Create an audit trail of modifications to the images and metadata.

- Control access to images.

- Provide an inventory of completed work.

- Automate and validate data entry.

- Facilitate use of controlled language.

- Provide a home for the metadata needed for long-term access to the images.[1]

An image management system is essential to the success of any digital imaging project. No single image management system solution, however, is likely to be appropriate for all digital imaging pro-

---

1. John Weise, "Overview of Media Database Issues and Technologies," *OIT Digital Media Solutions*, www.oit.itd.umich.edu/projects/DMI/answers/pickmediadb.html. Howard Besser and Jennifer Trant, "Introduction to Imaging: Issues in Constructing an Image Database," *Getty Art History Information Program*, www.getty.edu/gri/standard/introimages/index.html.

jects. As with all electronic information management systems, the design and implementation of any individual system depend on the specific requirements of each particular implementation.

Solutions vary greatly in complexity, performance, and cost. The ideal solution for any particular implementation considers the purpose, audience, and function of the system. In many cases, a combination of systems may meet your needs. For example, it is quite common to have one system for in-house use but a different system for public access that uses a subset of the metadata.

USES OF THE SYSTEM

When selecting an image management system, first consider how the system will be used, including:

- ▸ Purpose
- ▸ Size
- ▸ Complexity and volatility of the data
- ▸ Generic database management functions, including authentication, audit requirements, system monitoring, and backups
- ▸ Expected demand and performance
- ▸ Available technical infrastructure—hardware, software, and personnel
- ▸ Cost

Purpose

Start by considering the purpose for the system. It is impossible to know what image management system you should select or build if you do not know what you want it to do. Will it be a public system, or is it intended only for in-house use? Will the database be static or will staff need to be able to update and modify records? If so, do you need to track who did this work and when? Will users need special tools for manipulating images, annotating them, or otherwise working with the raw material of the database?

Size

Consider the number and size of your digital resources. A system that consists of tens of thou-

sands of images might be best managed with a relatively simple and inexpensive desktop database system. A system containing hundreds of thousands of images may need more robust software. Similarly, a simple desktop solution might well accommodate relatively brief records, whereas powerful relational or object-oriented databases may be needed to manage extensive or very complex metadata.

Benchmarking according to the size of the database is one of the most difficult tasks in selecting an image management system. You must balance the number of records, their size, the number of simultaneous users, the capability of the hardware running the software, and the desired performance of the entire system. You must also take into account your organization's ability to develop and maintain a complex database. Given that all of these factors are likely to change, it is wisest not to lock yourself into a limited solution. Prototype a small system that is easy to design and implement, but make sure that you can easily move the image management system to a more powerful and complex environment if demand or system requirements dictate it.

Complexity and Volatility of the Data

Image management systems may have data that is relatively simple, such as just a title and a file location. Or they may become very complex with extensive information about the description, history, and use of the asset represented by the record in the database. A complex database may have a thousand or more fields linked by hundreds of tables.

Similarly, the resources managed by an image management system may be static. You may wish to digitize a single collection, for example. Having scanned and identified the images, you may not need to further update or modify the image management system, the descriptive information and the file locations of the images being unlikely to change. A different project might involve volatile metadata or an ever increasing number of images. The image management system requirements for these two scenarios differ a great deal.

If all you require of the database is simple keyword retrieval and file location, it makes little sense to acquire and maintain the most complicated database system in the world. Conversely, don't expect a spreadsheet or a word-processing package that mimics a database to provide all the features that are found in full-fledged database packages. While it might be possible to mimic the authentication, data normalization, audit capabilities, and other features of a powerful database with a small system, your programming costs will be high and performance will suffer.

Authentication and Audit Requirements

In complex systems it is often necessary to know both who is performing certain functions (such as adding, modifying, or deleting records) and that they are authorized to do so. It is therefore important to specify your security requirements. You might require users to log into the system with a user name and password, for example, or require permission from an institution-wide authentication server. Similarly, different classes of users might be able to view images at different sizes. Concerns about copyright, for example, might lead a university to provide large images to students enrolled in a class, smaller images to other campus users, and still smaller images—or none at all—to the general public. Specify any system security requirements in advance to include them in the image management system design.

Expected Demand and Performance

How popular will your image system be? As demand on the system increases, performance usually drops. Before selecting a system, determine how many people will need to use your image management system concurrently and how intensively. A system that is accessible only to a few staff members working on-site at any one time will not have the same hardware and software requirements as a system that is open to the entire institution or, if on the Web, to the world. Also, determine how responsive you feel the system should be. Must a search of the database seem almost instantaneous, or is making the user wait for several seconds (or even a minute) acceptable? You can then choose between a simple desktop system or a more robust system running on a server.

When calculating demand, remember that it may fluctuate, and decide whether to build for average or peak usage. UTOPIA, a pilot database

of Renaissance art for students at Cornell University, did not anticipate changes in demand. Initially, the system was properly sized for demand during the semester but searches slowed to a crawl and several times the system crashed under the end-of-semester load. It has since been moved to a more robust platform. Of course, it is often difficult to anticipate demand. When the Church of Jesus Christ of Latter-day Saints offered free access to many of its genealogical databases, demand far exceeded expectations. The site had been built to handle 25 million hits a day—a huge load by any measure—but in the first few weeks after it opened to the public, it registered at least 40 million hits a day and turned away another estimated 60 million hits a day.[2]

All information systems should face the problem that they are too popular. A more common failing is estimating too much demand. This leads to building a system with an infrastructure that is more complex than is necessary and is difficult to maintain. A limited but scalable design is usually the best approach.

Available Technical Infrastructure—Hardware, Software, and Personnel

Design according to the technical infrastructure available to you.[3] The design phase is also a good time to investigate potential infrastructure collaborations, either with parallel or overarching organizations. Such infrastructure collaborations can reduce cost burdens, and also lead to explorations of metadata relationships that over time bring together intellectual access to collections.

No technical resource is more important than a skilled staff. Most current digital imaging systems are built from a combination of commercial off-

2. "New Family Tree Internet Service Swamped by Demand" (May 26, 1999), *Church of Jesus Christ of Latter-day Saints, News Updates*, www.lds.org/med_inf/new_upd/19990526_FIGS_ Demand.html.

3. The CPA/RLG Task Force on Archiving of Digital Information identified the need for a deep infrastructure for the successful archiving of digital information, but the need is broader than preservation and should be part of any digital imaging program. Task Force on Archiving of Digital Information, *Preserving Digital Information: Report of the Task Force on Archiving of Digital Information* (Washington, DC: Commission on Preservation and Access, 1996); also available as "Preserving Digital Information: Final Report and Recommendations" (May 20, 1996), *RLG*, www.rlg.org/ArchTF/index.html.

the-shelf products and custom programs. No one turnkey system meets all system needs in all areas, especially user interfaces, metadata, and migration strategies. Skilled staff integrate all the component parts of the imaging system and keep it running.

As Bruce Bruemmer demonstrates in his sidebar, even supposedly complete commercial systems still require extensive in-house support. Perhaps more surprisingly, small-scale systems developed in-house can also require a high level of support. It is quite easy to develop, for example, an image database in a standard small computer database package such as Microsoft® Access or FileMaker® Pro. Many repositories have done so, often relying on student help or the skills of a technically savvy staff member. But when that student or staff member leaves, can the institution guarantee the staff and the skills to maintain and upgrade the system? And how will the system be monitored to ensure that it operates continually? Frequently the simple, seemingly inexpensive staff-built system becomes the most costly to maintain. An investment in technical expertise at the start of the program may actually result in savings later on.

Cost

Take the overall cost of the system into account when you select an image management system. Unfortunately, it is usually impossible to assess the value of the benefits that may be gained from implementing an image management system. A good but a more costly system, for example, may be able to reduce staff time spent on reference work or speed the reproduction ordering process. However, no literature suggests that the savings you can realize with a good image management system surpass the expense of acquiring and maintaining the system. Each institution, therefore, must determine how much money to spend to improve access.

SYSTEM DESIGN FEATURES

Consider general system design features and how they may contribute to efficient use of the system, including:

▸ Flexibility

▸ Single versus multiple databases

▸ Adherence to data system, structure, and content standards

▸ Expected life span of the data and potential for migration[4]

Flexibility

Given the rapid evolution of the universe of image applications, the design of any image management system you select today will most likely change in

---

On the Bleeding Edge
Bruce H. Bruemmer
Coordinator, Digital Collections
University of Minnesota Libraries

In 1997, the University of Minnesota Libraries had not yet established a digital library program, although it seemed that every other research library in the country had. Then, Minnesota's new university president announced that he wanted to make the University of Minnesota a leader in digital technology. Clearly, we were motivated to find a way to implement a program quickly.

Over several months, the university librarian invited major companies with expertise in imaging and document management to propose solutions to get the library started. The presentations involved some major corporations, but none offered turnkey systems that seemed at all related to the needs of a research library. The most promising discussion was with IBM, which sold a software product named the IBM Digital Library. The version I software ran only on an AIX system (IBM's version of UNIX) and employed either the company's stalwart DB2® database software, or Oracle® software. We invited the IBM sales staff to inspect a likely special collection and they suggested a number of applications, from complete reproductions of books to streaming video. While no specific library application was demonstrated using the software, a number of respectable clients were mentioned, foremost among them the Vatican. Software selection is one of the most difficult and imprecise of modern human endeavors and any air of infallibility was a welcome breeze. We were looking at the world's oldest computer firm, a stable system, and a list of important clients. An independent software review firm noted that there were no other competitors in this area. How could we go wrong?

CONTINUED ON PAGE 123

---

4. Drawn in part from Lois F. Lunin, "The Big Picture: Selection and Design Issues for Image Information Systems," in *Digital Image Access & Retrieval*, ed. P. Bryan Heidorn and Beth Sandore (Urbana, IL: Graduate School of Library and Information Science, University of Illinois at Urbana-Champaign, 1997), p. 43.

the not-too-distant future. For the present, consider even the best image management system to be a work in progress. Successful image management systems tend to grow beyond the wildest expectations of their founders, in both size and use. It is imperative, therefore, that a system be scalable so that if it succeeds, it can grow. Flexibility may be desirable even within a current, limited, application. For example, you may need a cross-platform solution or an image management system that is compatible with other databases.

Single Versus Multiple Databases

One simple way to ensure a flexible design is to develop multiple image management systems, each suited to a particular task. For example, a database that repository staff use for collection management, with information on the physical location of artifacts, how they were acquired, their condition, and the virtual location of their digital surrogates, may not be the best system for public access. A second image management system could include extracted data for public access. Harvard University adopted this approach for a project to develop a union catalog of visual image resources. The Visual Information Access (VIA) catalog serves "as a central place where researchers can discover the wealth of visual materials available across the university and learn where at Harvard they can go for access to the materials themselves or for more detailed information about them." The data in the public access system comes from local systems that "will always contain richer information and provide capabilities tailored to local needs."[5] Image Services at the University of Michigan's Digital Library Production Service has implemented an image access system that combines content from disparate image collections without compromising their richness. Users can search collections in the system in any combination by generalized data categories. Additionally, each collection in the system is searchable by collection-specific attributes. The system currently provides access to museum, library, archive, special collections, and visual resources data, primarily from campus units.[6]

5. "Database Scope," *Harvard University, Library Digital Initiative*, hul.harvard.edu/ldi/html/database_scope.html.
6. "Image Services," *University of Michigan, DLPS*, images.umdl.umich.edu.

Implementing multiple databases can offer real advantages. Each system can be optimized for its particular use, the separation of the two systems can provide security (corruption of the public system does not affect the internal system), and excessive demand in one system does not adversely affect performance in the second.

Adherence to Data System, Structure, and Content Standards

Data system, structure, and content standards ensure flexibility. They can make it easier to exchange information between systems and aid in the migration of data.

CONTINUED FROM PAGE 122

A small group of librarians formed to add some reality and practicality to the expectations of the pilot project. Our goals were still ambiguous, although shaped by the notion of delivering rare images on the Web, preserving documents by reducing use of the originals, avoiding copyright pitfalls, and inspiring other digital projects at the university. The group chose the Archie Givens, Sr., Collection of African-American Literature, which was richest in 20th-century books. The project focused on graphics within the literature collection, primarily from book jackets. The jackets were rare, some had preservation problems, there was a demand for them, and many contained stunning images. The emphasis on book jackets meant that the software would need to be configured to bundle the images together with one descriptive record and a file derived from OCR applied to the jackets. Over the next three months we fashioned a data model with one of IBM's business partners.

After installation of the system, reality hit hard. The system was flexible, but required a programmer in IBM's special macro language. There was no interactive loader for images and data, simply a batch program that relied on tagged and formatted data. Worst of all, the keyword-indexing program prevented editing or deleting data already entered into the system. Software developers at IBM *thought* some of these problems (and others!) could be fixed, but no quick consensus emerged. We had a potential disaster on our hands and we were seriously considering opting out.

Five months after the purchase, Ancept Inc., a small Minneapolis software firm that specialized in the IBM Digital Library, contacted the library about the project. They received a full briefing about our woes, and offered to help. They already had some code that they thought would improve the software and were building a relationship with IBM. After the library negotiated with IBM, aided in part by our IBM salesman,

CONTINUED ON PAGE 124

CONTINUED FROM PAGE 123

Ancept was given the go-ahead to work with the library to improve the usability of the system.

Over six more months, a system emerged that met our basic expectations. At the beginning of the process, no one anticipated the number of phone calls, e-mails, meetings, cajoling, and heartburn that this project would require. We became a good deal more realistic about future projects and learned the following the hard way:

- Archivists and librarians understand indexing and intellectual access far beyond the capabilities of commercial software. A software developer or marketer making five times the salary of a librarian is likely to be a bit arrogant about a librarian's understanding. Do not sell yourself short in what you know. We assumed that something named the Digital Library would be well informed by library practice. Assume nothing. What is obvious to an information specialist is probably novel to a software developer.

- Language and cultural differences among librarians, marketing staff, software developers make communication even more problematic. What was a field to us was an attribute to the software people. Our database schema was their data model. And what was a little bit of programming to the developers was something we had no staff to conduct.

- Be specific and precise about the goals of your digital project or have a fallback plan. This advice seems overly simple, but with so many exuberant expectations about digital libraries it is difficult to confine a digital project to straightforward goals. Had we begun with clearly defined expectations, we would have identified issues with the IBM Digital Library prior to purchase. At minimum, client and vendor alike would have better understood the performance goals of the software.

- Give your vendor a chance to do right. In our case, IBM recognized that it had left an important customer in the lurch. Its primary salesman to the university recognized and was concerned about the library's unhappiness, and he became an important advocate within the corporation to make the situation workable. Keep cool, but resolute.

- Be honest about your needs for information technology support, especially in an academic environment. Have you put the success of your project in the hands of a server administrator who is a student and likely to leave in three months? Can you support the customized programming that is inevitable even with a turnkey system?

- Pretend you are traveling in a foreign country and don't know the language or the local customs. In other words, be prepared for the unknown. If you choose to establish a partnership with a company that has untested software, you will encounter surprises and disappointments. If your goals are ambiguous and you use imprecise language, be prepared for a long, indirect, scenic ride.

The project Web page is located at brainiac.lib.umn.edu.

- *Data system standards*: Data system standards and protocols can facilitate interaction between systems, almost without regard to the specific content. It is easier for two databases to connect, for example, if they both comply with the Open Database Connectivity (ODBC) specification. It may be easier to update or modify data if it can be accessed using Structured Query Language (SQL), a database query language adopted as an industry standard in 1986. And it would probably be easier to integrate an image management system with other databases (including bibliographic databases) if it complies with Z39.50, the *Information Retrieval: Application Service Definition and Protocol Specification*, ANSI/NISO Z39.50 and ISO 23950.[7]

- *Data semantics and structure standards*: MARC is probably the data structure standard most familiar to librarians and archivists. For certain kinds of databases (primarily those with one record for each image), the MARC format may be appropriate. At a minimum, you may wish to specify MARC as an export format to assist in the migration of the database.[8] Structures defined by SGML can also be valuable for migration and metadata preservation.

Other communities have adopted methods and syntaxes for metadata that may be appropriate for an image database. Many newspaper photograph systems, for example, have adopted the Information Interchange Model (IIM) of the International Press Telecommunications Council. The model specifies a format for a news object container file and includes the full specification for editorial metadata. The University of California at Berkeley has recently proposed a standard to express the structure of scanned

7. For information on ODBC, see the Microsoft ODBC home page, *Microsoft Universal Data Access Web Site*, www.microsoft.com/data/odbc. A good starting point for SQL is "SQL FAQ: Table of Contents," *Sequoia 2000 Illustra Frequently Asked Questions*, epoch.cs.berkeley.edu:8000/sequoia/dba/montage/FAQ/SQL_TOC.html. For information on the status of the standard, see "JCC's SQL Std. Page," *JCC Consulting, Inc.*, www.jcc.com/SQLPages/jccs_sql.htm. On Z39.50, see "Library of Congress Maintenance Agency Page for International Standard Z39.50," *Library of Congress*, lcweb.loc.gov/z3950/agency.

8. On MARC, see Library of Congress, Network Development and MARC Standards Office, *MARC Standards*, lcweb.loc.gov/marc.

archival documents. Art slide librarians have developed the Core Categories for Visual Resources, an attempt to express the data structure they feel is needed to describe visual documents depicting art, architecture, and artifacts. The Consortium for the Interchange of Museum Information (CIMI) has proposed using the Dublin Core, an emerging standard for the description of Web-based objects, as an organizing principle for the interchange of museum information. Archivists have adopted a standard expressed in SGML entitled Encoded Archival Description (EAD) to represent the organization of information in archival finding aids. Geospatial systems, space sciences, and medical imaging all have their own data structure standards.[9]

▸ *Data content standards, including access to a controlled thesaurus*: Data content standards control the content inserted into the pigeonholes defined by data structure and semantic standards. Data content standards enable consistent access to images and in conjunction with data structure standards contribute to the data's longevity. Consistent, standard terms enhance the system. Using the same terms to describe the same types of objects enhances searching across systems. And standardized use of content makes data migration easier. You can, for example, move data relatively easily from a system that has a single field for "name" to a system that uses "first name" and "last name" if data in the original system is consistently entered as "last name, first name." Automated tools could in this case parse the data from one into two fields. Many standards apply to data content in image management systems. Images are often described using standardized thesauri such as the Art and Architecture Thesaurus, ICONCLASS, or the Library of Congress Subject Headings.

9. On the IIM, see the International Press Telecommunications Council home page, www.iptc.org/iptc. The Berkeley data structure model is detailed at "The Making of America II," sunsite.berkeley.edu/moa2. For the VRA Core Categories, see "Data Standards," *Visual Resources Association*, www.oberlin.edu/~art/vra/dsc.html. For CIMI's Dublin Core work and the Dublin Core project, see the CIMI home page, www.cimi.org, and the Dublin Core Metadata Initiative home page, purl.oclc.org/dc. For more on EAD, see Library of Congress, Network Development & MARC Standards Office, *Encoded Archival Description Official Web Site*, lcweb.loc.gov/ead. For an excellent general guide to many of the metadata systems in use, see "Digital Libraries: Metadata Resources," *IFLANET*, www.ifla.org/II/metadata.htm.

Expected Life Span of the Data and Potential for Migration

Librarians and archivists tend to assume that the data systems they develop are permanent. They therefore seek to design systems that will be of enduring use. In certain cases, however, systems that by design have a limited life span may be more advantageous. Public access kiosks, for example, or reproduction order systems might be designed to fill a specific, limited need, after which time they could be discarded. For these applications, proprietary systems that rely on unknown data types and file formats may be perfectly acceptable. In most cases, however, libraries and archives will wish to maintain, reuse, or repurpose data developed for specific applications. The image management system must be able to migrate data to successor systems.

There are several ways to ensure that the metadata in an image management system today will continue to be available and of use in the future.[10] Ideally, design every system with a clear migration path for the data. Current data system, semantic, and structure standards (such as SQL, or MARC, or SGML) make it easier to retrieve and manipulate data in the future. Even a nonstandard database should have some means of exporting data in a delimited, usable format, so data should follow a data content standard. Unstructured, uncontrolled, inconsistent data, even if it is in the most open system imaginable, is unlikely to be of future use. Adherence to either community metadata standards or, at a minimum, internally consistent data standards remains the surest guarantee of future data viability.

IMAGE MANAGEMENT SYSTEMS IN PRACTICE

Myriad options are available for image management systems. Most of them, however, can be grouped into a few generic approaches, including:

▸ In-house systems using common desktop databases

▸ In-house systems using client/server architecture

10. Other chapters discuss steps to ensure that image file formats remain of use; this chapter considers only metadata.

- Specialized image management programs designed to run on the desktop or on a client/server architecture

- SGML/XML-based solutions

In-House Systems Using Common Desktop Databases

A number of powerful database programs can run on a desktop computer as well as on a small server. The Microsoft Access, FileMaker Pro, and Corel® Paradox® programs are the three most commonly found at the heart of homegrown image management systems. Most desktop systems can be used for either relational database management systems (RDBMS) or for flat-file systems. Relational systems normally enjoy advantages in performance and complexity over flat-file systems. Flat-file systems are often easier to design initially, however, and it is possible to design attractive, useful applications that employ a flat data structure.[11]

Advantages

Developing an image management system using one of these programs offers great advantages. For one, there is usually a low initial cost for the desktop programs. They are often included in suites of software, or cost a few hundred dollars at most. Because they are intended for the nonprogrammer, it is often relatively easy to design and implement applications using the tools that come with the software; they do not require knowledge of a programming language. Because they are relatively simple, it is often easy to import and export data from the applications. And finally, because they are so common, extensive suites of commercial tools and networks of user groups can assist in the design and use of the programs. Many people, for example, are developing sophisticated Web interfaces to Microsoft Access databases using either Allaire's ColdFusion Web application server or Macromedia Drumbeat 2000™.

Disadvantages

The disadvantages of developing an image management system using one of these products are

also great. First and foremost, you have to design a system. Each of these programs provides the raw materials, but the design and construction of the finished system is left up to you. Some users and user groups have attempted to share their development experience. Susan Williams of Yale University, for example, has developed a generic FileMaker template for slide librarians, and archivists have established a mailing list to share their experience using Microsoft Access.[12]

Desktop solutions may also offer limited robustness. Sizing database solutions is an arcane and inexact art. How well a database will perform depends on the size of its server, the number and complexity of records, the presence of index files, and the number of simultaneous users. Anecdotal reports indicate that desktop systems have become unbearably slow with as few as 10,000 records or with a dozen simultaneous users, though other applications run successfully with 100,000 records or more.

The ability to rapidly develop and implement an application using a desktop system is appealing. The difficulty in estimating the performance of such a system makes it all the more imperative that the application be designed so that in the event that performance suffers, the data and images can be moved to a more robust environment.

One final difficulty with desktop management systems arises precisely from their ease of use. Almost anyone can sit down and design a database using Microsoft Access or FileMaker Pro software, but few people either know how to design a database well or document what they have done. Too often, when designed by a knowledgeable student staff member, the initial low investment in programming leads to high costs down the road, when someone else has to figure out how to modify, migrate, or fix a broken or obsolete program.

In-House Systems Using a Client/Server Architecture

A step above the desktop systems in performance but also complexity are true client/server database applications. At a minimum, a true client/server application separates the database and its associ-

---

11. A good example of a flat FileMaker® Pro database linked to Web pages is *NYS College of Human Ecology—Historical Photographs*, rmc.library.cornell.edu/edb-hephotos.

12. Direct subscription requests for MSACCESS4ARCHIVES to listproc@lists.princeton.edu. Susan Williams's FileMaker template: www.library.yale.edu/~swilliam/fmp.htm.

ated activities such as searching or data authentication from the presentation of the information on the client. Commercial—and frequently expensive—RDBMS products in this arena include programs such as 4th Dimension from ACI US and others from Oracle, Informix, and Sybase. The MySQL database server by T.c.X can be free, but may be less robust.

Object-oriented, as opposed to relational, databases represent a second kind of client/server database.[13] Many of the desired features in an image management system, including the hierarchical relationship of data and the ability to index and manipulate many kinds of digital files—text, image, sound, and video—lend themselves to object-oriented databases. Unfortunately, pure object-oriented databases from companies such as Novell, Objectivity, Inc., Object Design, Inc., and Computer Associates have yet to establish themselves as a significant component of the database market and their potential is by and large unrealized.

While client/server database management systems may offer better performance than desktop systems, they also share many of the same problems. Developing a system, from formal data modeling and data architecture through actual coding, implementation, and maintenance, is expensive. All but the largest consumers normally turn to commercial products that have been developed around the core engines.

Specialized Desktop or Client/Server Systems

A third category of software consists of commercial off-the-shelf software designed with image management in mind. Such systems include simple programs designed to help the home user manage scanned photographs formerly kept in shoe boxes, desktop image management systems designed to manage generic collections of scanned images, and client/server applications developed for specialized communities including general and digital libraries, museum management, media management (often involving video), document management, and

other specialized uses. While none are ideal, many can be adapted to meet immediate needs.[14]

Desktop Systems

For many small repositories, especially those interested in managing photographic collections, one of the desktop systems might be the best solution. Among the most widely used are Canto Cumulus®, ImageAXS Pro from Digital Arts & Sciences Corporation, and Extensis® Portfolio™ (formerly Fetch).[15] Desktop commercial image management programs have many advantages. Most have a predefined data structure, so you do not need to develop or design your own template. Most incorporate built-in links to images from a textual metadata record. Some are cross-platform, allowing access from the Microsoft® Windows® operating system and Macintosh® machines. Because these are off-the-shelf products designed to be run by a novice, they usually require less programming expertise than do programs that have to be configured.

In certain situations, the disadvantages of these programs can outweigh the benefits. For one, many have a fixed or comparatively inflexible data structure. While you do not have to design the structure, it is also hard to modify or change it. In at least one case (Canto Cumulus), the system is not based on fields, but instead provides a limited set of keywords. You can build hierarchies of keywords to emulate fields, but you lose much of the simplicity of the database. Some of these products are built around common database engines (such as the Microsoft Jet engine that underlies Microsoft Access software); others use proprietary engines, which may cause problems when it comes time to repair or migrate the database.

Client/Server Applications

For larger applications, libraries and archives will wish to turn to true client/server programs. A good starting point for many libraries may be their own library catalog. Many libraries are installing image-

---

13. Most of the major RDBMS vendors are in the process of incorporating support for objects in their systems to form a hybrid, often referred to as object-relational database management systems (ORDBMS).

14. For an excellent list of many media management programs, see *Footage.net: The Stock, Archival and News Footage Network*, www.footage.net.

15. Canto Software home page, www.canto-software.com; Digital Arts & Sciences/Gallery Systems home page, www.dascorp.com; "Extensis® Portfolio™ 4.0 for Asset Management," *Extensis*, www.extensis.com/portfolio.

Delivering the AMICO Library™ through RLG's Eureka® Service

Arnold Arcolio
Information Architect, Research Libraries Group
Bruce Washburn
Information Architect, Research Libraries Group

To provide access to the AMICO Library™, a joint digital library documenting museum collections assembled by the Art Museum Image Consortium, the Research Libraries Group expanded its Eureka® service to support AMICO's testbed year, from August 1998 to August 1999.[1] At that time, the AMICO Library represented nearly 20,000 works of art with item-level textual metadata and at least one digital image for each work. Some works are represented by several views, but none involve complicated sequential or hierarchical relationships. RLG worked with AMICO to develop suitable forms of access, presentation, and navigation, and useful ways to manage results.

Implementing RLG's Eureka for AMICO

RLG's Eureka service was already established as a powerful Web interface to scholarly resources for higher education, providing integrated access to materials from many different sources. Eureka's discovery mechanisms offer users various ways to find materials with which they may be unfamiliar:

- Separate simple and advanced search screens, including constantly visible, context-sensitive help for index scope, use guidelines, and examples.

- Result management: sorting, paging, jumping forward half way or to the end, reviewing and returning to previous search results.

- Keyword searching, word searching to compensate for forms in different order, and browsing of phrases to compensate for alphabetic variants that sort near each other.

The AMICO Library was made available to higher-education users for discovery and acquisition of images and related documentation, with uniform rights for educational use. We expected that the descriptive data would help users find images and that for many users the images themselves would be of primary interest for downloading and local study. Users should be able to:

- Find a specific work, works associated with a specific artist, and works of a specific material or type.

- Save results and individual works for later study.

- View images within the Web browser, without needing a plug-in for most uses.

- Download images and metadata.

- Access high-quality source images when necessary.

To make the AMICO Library useful, we extended the Eureka design to:

- Include thumbnail images on brief search results that show several different works and larger images on full displays of individual works.

- Provide display options for text-and-image, image-only, and text-only brief displays.

- Create separate display windows for comparing images.

- Offer different access points: keyword, creator, title, type, materials and technique, date, subject, owner name, owner place, accession number.

- Maintain a persistent "notebook" for storing information about works across sessions, with multiple notebooks available to individuals.

- Display rights information for every image and work, and links for institution-specific, commercial-use information.

- Include technical information for images (file size, dimensions, technical details of their creation).

- Allow ordering of source TIFF images.

- Ensure a secure path to images to prevent access from outside the service.

AMICO Features Enabled Effective Implementation

The Consortium provided a forum for discussion of practice among contributors and the AMICO Data Dictionary provided a data-structure standard.[2] The dictionary defines what elements can be present but does not provide rules for description or formulation of those elements or what forms those values should take. The set of core fields it defined helped us put together a functioning interface, since we could rely on their being present. The dictionary allowed for recording of display and indexing versions of some data (e.g., creator name, dates, dimensions) so we could display a name in the form "Winslow Homer" and provide a sorted list of names for browsing using the form "Homer, Winslow." The Related Image Description data element provided us with a label we could use anywhere in the interface to identify an image; for example, Detail or X-Ray. This allowed catalogers to control important text that identifies images at the highest level of the interface. The Consortium's ongoing efforts to parse and normalize date values like "8th Century BC" and "Before Mid-1880s" into numerical values suitable for searching and sorting should support better access and collection organization based on date ranges.

The convention of beginning image file names with a distinctive code for the contributing institution helped us manage image processing, allowed us to organize images for delivery through the interface, and simplified reporting of image use. This code was the only important feature of the image name, from our point of view, apart from its being unique.

CONTINUED ON PAGE 129

enabled library catalogs from vendors such as CARL, VTLS Inc., SIRSI, and Endeavor. In addition, as described by Arnold Arcolio and Bruce Washburn in their sidebar, RLG offers access to the AMICO Library™ of museum images through its Eureka® service. It is likely that in the near future all library systems will be able to deliver full text and images.[16]

Existing library catalogs have many advantages

16. For an interesting example of a photograph collection using the CARL system, see *Photography Collection, Denver Public Library, Western History/Genealogy Department*, gowest. coalliance.org. The Library of Virginia makes textual documents available through a library catalog using VTLS software as the basis for the system: "Digital Library Program," *The Library of Virginia*, www.lva.lib.va.us/dlp.

as image management systems. First, library catalogs represent a huge preexisting investment in metadata. It makes sense to try to exploit rather than reinvent this resource. Secondly, the emerging library catalogs permit ready links between the catalog metadata and digital images. In some cases, you move from text to image; in others the text includes at least a thumbnail. Finally, most library catalogs are maintained by a highly skilled technical staff. By piggybacking digital projects onto the existing software, imaging projects can take advantage of the existing expertise.

Although few, the disadvantages of using library catalogs as an image management system are significant. First, most catalogs still rely on the MARC format. As long as there is one MARC

CONTINUED FROM PAGE 128

The model of having contributors supply a high-quality TIFF image and having the distributor derive images suitable for delivery via the Web interface proved convenient, allowing for experimentation in appropriate dimensions and conversion utilities.[3] It allowed us to offer standard image sizes for all contributors without their having to understand and follow a specification from us. As technology improves, we will also be in a position to produce better derivative images without additional action on the part of the contributors.

We derived four sizes of images for display within the Eureka interface, all with minimal JPEG compression:[4]

▸ *Thumbnail*: 128 pixels maximum height and width, shown in brief multi-item displays.

▸ *Snapshot*: 250 pixels maximum height and width, for display beside full textual data.

▸ *Inspection*: 480 pixels maximum height, 640 pixels maximum width, for study and presentation.

▸ *Presentation*: 768 pixels maximum height, 1,024 pixels maximum width, for study and presentation.

Deriving images at the distribution point allowed contributors to focus on providing an image that is as good as it can be, or is as good as they choose to make it, separate from our decisions about suitable forms for delivery that might change more quickly.

Lessons Learned

▸ Users are eager to comment, through the interface, on what they hoped to find but did not. We expect to provide a tool in the interface that captures this information for contributors.

▸ Image files and descriptive text are often created by different processes and people. We found that image names in

the descriptive text and for the image files themselves sometimes differed slightly (in case or punctuation, for example). This would cause the image to be missed when the description was viewed through the interface. This is an area where a contributor's quality assurance efforts are very valuable to us as distributors.

▸ Seeing their various descriptive practices side by side in an integrated resource motivated contributors to work toward descriptive standards. Meanwhile, keyword searching and careful indexing can help users contend with that variety.

▸ We experimented with enhancing searching by creator using the Getty Research Institute's Union List of Artist Names. While not yet visible in Eureka during the testbed year, our research suggests that such tools will improve retrieval when used in conjunction with browsing and careful indexing for data that is not all created using a single controlled vocabulary.

We know that the utility of data is increased by collaboration on standards and practices among contributors. And it is now clear that the value of individual museum collections is greatly increased by integration. In addition, we believe our further integration of image resources with related resources in other forms (for example, slide library catalog records, and archival collection guides) will make digital collections even more accessible and meaningful.

Notes

1. *AMICO, Art Museum Image Consortium*, www.amico.net.

2. "Amico Data Specification," *AMICO*, www.amico.net/docs/dataspec.html.

3. The AMICO Related Image and Multimedia File Specification calls for 24-bit color and a minimum resolution of 1,024 by 768 pixels; most contributors met or exceeded this guideline.

4. Using ImageMagick™ 4.05 with JPEG compression quality level 90 for most images.

record for one image (as sometimes happens with photographic collections), the reliance on MARC seldom poses a problem. The biggest issue is that the MARC format may not easily accommodate some desired metadata, such as information on the scanning process or information about multiple derivatives. In addition, the mixture of item-level records in a bibliographic catalog can lead to problems associated with the granularity of information. Is it useful, for example, to retrieve references to 300 individual photographs of the Chrysler Building found in an architecture collection if you only want a book on the subject?

A museum-oriented collection management program offers a powerful alternative to the library model, especially for collections of primarily graphical material. While not image management programs per se, like library catalogs, many of these programs can handle images and can fill the role of an image management system. Among the most common are Willoughby's Multi MIMSY 2000, EmbARK™ by Gallery Systems, ARGUS by Questor Systems, and STAR® from Cuadra Associates (a flexible program that is often used for libraries and records management).

While perhaps not quite as developed a field as library management software, museum management software systems boast several advantages. Many accommodate complex predefined data structures and use sophisticated database engines. Most offer clients for both Windows and Macintosh platforms.

Conversely, the fixed and complex data structure can impede simple image management. Some—most notably Cuadra's STAR—are based on a nonstandard underlying database engine, complicating interaction with other systems. These products, like the library systems with which they compete and in comparison to the desktop systems, can be expensive and difficult to maintain and support.

A third possibility, especially of interest for primarily textual collections, is a document management program. Developed to facilitate the use of scanned textual documents, these software packages allow staff in insurance companies or banks to get access to the documents in a case file, for example, or to scan and store all incoming correspondence. Examples include Panagon™ by FileNET, Domino.doc™ by Lotus, and Folio™ and LivePublish™, text-based approaches to document man-

agement from NextPage, LC.[17] These large, sophisticated programs can handle hundreds of thousands of pages of text. Most focus on textual documents, though multimedia objects are slowly being added. Unfortunately, their data elements usually support the business market, not libraries or archives. Few, if any, are even designed for records management requirements, making it difficult to replicate the complex hierarchical arrangements libraries and archives use for information. The systems are usually proprietary rather than open and are almost all expensive. Still, for a large textual archive, document management systems may be a real option. Deborah Skaggs's sidebar discusses the strengths and weaknesses of document management systems.[18]

SGML/XML-Based Solutions

A relatively new approach to image management uses either an SGML or XML Document Type Definition (DTD). Not themselves software programs, SGML and XML DTDs describe the component fields of an image management system. Users search the encoded text with a software application, most frequently a textual search engine such as XPAT (originally distributed by Open Text as OT5™ and now available from the University of Michigan) or DynaText® from Inso Corporation. The encoded text may also be stored in a database and searched as if it were any other data in fields. For SGML-encoded data, a server converts the results to HTML for display to users, since most users do not have the software needed to display native SGML documents. Some XML documents can currently be viewed directly in Microsoft® Internet Explorer

---

17. An interesting example of document management systems at work is *TobaccoResolution.com*, www.tobaccoresolution.com. Here two different document management systems (from FileNET Corp. and Basis International) and one search engine (Alta Vista) are used to index and deliver the thousands of pages of tobacco-company documents released to the public as part of the settlement with the state of Minnesota.

18. I know of no implementation of document management solutions as part of a library or archive delivery system. Initial plans by the Center for Retrospective Digitization, Göttingen State- and University Library to use the FileNET product at the heart of their document management system collapsed because of "the complex requirements of library materials and the demand for open and platform-independent metadata and document structures." "Center for Retrospective Digitization, Göttingen State and University Library," *Göttinger Digitalizierungs-Zentrum*, www.sub.uni-goettingen.de/gdz/en/gdz_main_en.html.

5 and the capability has been promised for the next release of Netscape's browser.

The success of SGML and XML depends on the DTD employed. DTDs for several different kinds of image materials have been adopted, but none widely.[19] The Electronic Binding Project, or Ebind,

DTD developed by Alvin Pollock and Daniel Pitti at the University of California at Berkeley delivers page images of textual materials. As the Ebind Web site notes,

The Ebind SGML file records the bibliographic information associated with the document in an ebind header, the structural hierarchy of the document (e.g., parts, chapters, sections), its native pagination, textual transcriptions of the pages themselves, as well as

---

19. TEI, the Text Encoding Initiative DTD, has of course received wide adoption for the markup of textual materials in machine-readable form.

Electronic Document Management Systems as Archival Image Repositories

Deborah Skaggs
Manager of Corporate Records, Frank Russell Company

Today's electronic document management systems (EDMS) support the life cycle management of document-based objects—images, data, text, graphics, voice, and video. EDMS functions include the capture, storage, classification, indexing, versioning, maintenance, use, security, and retention of document-based information.[1]

EDMS products were first developed in the late 1980s as stand-alone solutions for specific applications, such as imaging, workflow, document management, or computer output to laser disc (COLD). Although these products streamlined business processes, improved user productivity, and enhanced enterprise-wide information access, they were not fully integrated. Typically, separate file servers or repositories for each application were not linked by a common interface, which meant that users had to execute multiple searches to locate a particular document.

Over the past several years, the consolidation of document management vendors and tools has promoted further integration. Today, many EDMS provide a common user interface—the desktop—that supports a single point of entry to documents without regard to their location or format and provides a common back-end architecture that supports centralized administration. A critical aspect of this single point of entry is the consistent use of document properties (i.e., index terms) for query and retrieval. The common back-end architecture means that only one system must be managed. This high level of integration also permits users anywhere in the enterprise to access information stored in the EDMS repository via a Web browser. In addition, several vendors are implementing records-management services, which support the disposition of documents and records. However, no EDMS currently offers preservation services to ensure the authenticity of documents over time.

Today's integrated EDMS have these features and characteristics:

- Web-based access and scalability

- Distributed architecture that is faster and more powerful than client/server architecture

- Customizable document architecture services such as document properties, classification, and access control

- Enterprise access support

- Tracking throughout the documents' life cycle, managing a range of content created in various applications, such as e-mail, databases, graphics, and images

- Foundation technology for an enterprise-wide information infrastructure that links content management and Web publishing

Advantages and Disadvantages of Implementing a State-of-the-Art EDMS

Advantages:

- Integrated management of multimedia library and archival material on a single technology platform

- Global access and mass distribution of digital images

- Safeguarded integrity and reliability of digital images

- Technical standards for system interoperability and digital image portability

- Controlled access to and improved navigability of digital images

- Reuse of information content of digital images, where appropriate

- Stability that ensures successful migration and reduces the risks of technical obsolescence

- Availability of work-management tools, such as automated workflow for internal business activities

Disadvantages:

- Needed creation or redesign of an information technology infrastructure and document architecture (e.g., document classification and indexing properties) to fully exploit EDMS functionalities

- A substantial investment in change management that

CONTINUED ON PAGE 132

CONTINUED FROM PAGE 131

includes redesigning business processes and promoting user acceptance of enterprise-wide sharing of documents.

▸ Costs of importing existing image database and index pointers to an EDMS

▸ Design and construction of an EDMS interface to an existing information/image-retrieval system

Comparing the pros and cons of using an EDMS may suggest that the benefits outweigh the disadvantages. Unfortunately, this may not be the case for several reasons:

▸ The initial and continuing investment of essential staff and financial resources is substantial.

▸ The redesign of an information technology infrastructure and document architecture involves specialized skills and experience that relatively few information technology specialists possess, to say nothing of archivists and librarians

▸ The design, construction, and validation of an EDMS interface to an existing information/image-retrieval system, particularly a legacy system, is formidable and typically involves experience and complex integration tools.

▸ The current costs of implementation are substantial and in most instances can be justified only by a large volume of images, sufficient revenue generation or savings, or substantial added value such as timely access.

▸ The absence of preservation services to ensure the authenticity of documents over time is a major limitation.

▸ Several EDMS services, such as ease of navigation and Web-based access, are already available as stand-alone applications.

Is the Use of an EDMS Justified?

Unless your objective is broader than the management of digital images, the use of an EDMS is problematic. The use of an EDMS to support the management of digital images could be considered in three instances. First, when an EDMS has been deployed that stores digital images created and captured as part of operational business activities. The EDMS could be extended to the archives and libraries to support the continued management of the digital images. Second, when there is a top-down commitment to deploy an enterprise-wide EDMS and the management of digital images in an archive or library can be incorporated into it. Third, when sufficient resources are available to support a customized EDMS application for use in a library environment.

Note

1. For more information on EDMS products, see Doculabs, Inc. home page, www.doculabs.com; FileNET Corporation home page, www.filenet.com; Documentum, Inc. home page, www.documentum.com; Hummingbird home page, www.hummingbird.com; *OnBase® by Hyland Software*, www.onbase.com; Identitech Inc. home page, www.identitech.com; Optical Image Technology, Inc. home page, www.opticaltech.com.

optional meta-information such as controlled access points (subjects, personal, corporate, and geographic names) and abstracts which can be provided all the way down to the level of the individual page.[20]

The Center for Retrospective Digitization in the Göttingen University Library has a slightly different approach. Called Agora, it relies on generating metadata in XML from a database. A similar experimental project at Cornell University to scan and encode law journals similarly uses well-formed XML to express the basic structure of the documents.[21]

SGML has also been used successfully to encode photographic and other image collections, most notably in the California Heritage Digital Image Access Project. In this project, the Encoded Archival Description DTD developed by archivists to encode finding aids to collections organizes access to photographic resources. The Image Services access system developed by the Digital Library Production Service at the University of Michigan uses a single SGML DTD to encode data in fields from a wide variety of image databases. The DTD supports the diversity of each collection database, and also enables cross-collection searching. The Museums & Encoded Archival Description (MUS-EAD) and the Museums and the Online Archive of California (MOAC) projects both investigated using EAD to describe museum objects. And most recently, in the Making of America II project, UC Berkeley has led a multi-institutional effort to develop a DTD that encodes the various metadata elements found in typical special collection materials.[22]

20. "Digital Page Imaging and SGML: An Introduction to the Electronic Binding DTD (Ebind)," *Berkeley Digital Library SunSITE*, sunsite.berkeley.edu/Ebind.

21. On Agora, see Norbert Lossau and Frank Klaproth, "Digitization Efforts at the Center for Retrospective Digitization, Göttingen University Library," *RLG DigiNews* 3, no. 1 (February 15, 1999), www.rlg.org/preserv/diginews/diginews3-1.html. For a demonstration version of the Cornell law journal system, see "Hein Online," brolly.cit.cornell.edu/Hunter/webdocs/Intro.html.

22. For the California Heritage project and the 28,000 photographs available through it, see University of California, Berkeley, The Bancroft Library, "Digitizing the Collection: Overview," *California Heritage Collection*, sunsite.berkeley.edu/CalHeritage/digital.html. The University of Michigan's Image Services system (images.umdl.umich.edu) will be formally available for local implementation in 2000; send e-mail to dlps-is@umich.edu for more information. For information about MUS-EAD, see "MUS-EAD: Museums & the Encoded Archival Description," *Museum*

There are many good reasons to consider an SGML or XML approach to organizing an image management system. For one, both are based on international standards and thus are inherently open in design. Because they can link to other SGML- or XML-encoded resources as well as to image files, it may be possible to build virtual collections drawn from multiple institutions.

Yet, there are also still many questions about the efficacy of SGML and XML as image management systems. First, SGML- and XML-encoded documents lack native client support, though given the general industry interest in XML this should change soon. Second, it is not at all clear that SGML search engines, which are optimized for large amounts of text, are as powerful as traditional relational databases for managing information. Agora relies on IBM's DB2® Universal Database software as the basic repository for its XML-encoded information, but not all information that one wishes to index in an image management system may easily conform to the confines of a database. SGML and XML systems do not as yet have many of the desirable features for databases, including data validation, links to authority files, and audit control. While the systems may work well for data and image delivery, they may not be optimized for data creation. Finally, while SGML and XML themselves are standards, there is still no agreement as to what data need to be retained in an image management system. In sum, adopting the SGML/XML solution presents many of the same advantages and disadvantages as designing your own database.

WEB DELIVERY

Most likely, even if your image management system is designed primarily for in-house use, you will want access to it via the Web. The quality of the Web interface in many ways determines the ultimate success of the image management system. Good data structure and content are of little use with a poor interface.

The Web displays HTML-encoded files. Query-

ing and presenting the results from an image management system via the Web involves connecting an HTML page in the browser, a Web server, and the underlying database. How you choose to make these connections depends on your database software, the availability and skill of programming staff, and the complexity of your database. The simplest way to deliver information from a Web database is to publish the data as static Web pages. Microsoft Access software, for example, and most of the specialized image management products offer static publishing of Web pages as an option. Web-server indexing software can then index the static pages.

If the data in the database program changes frequently, or if users need to be able to identify individual images or pages, or if there are large numbers of records in the database, static HTML files will not do. The most common way to implement a dynamic, interactive interface is to use CGI scripts, which are small programs that run on the Web server. In a typical scenario, a user enters data in an HTML form. The CGI program passes the information from the HTML form to a database, then wraps results from the database in HTML code to pass back to the user's Web browser.

CGI scripts are extremely powerful and versatile and are often the first choice for small applications. Unfortunately, they do not scale well. Since they run on the server, they use up server resources. And because CGI programs are usually written in a scripting language such as PERL, they do not use system resources as efficiently as do true programs written in a programming language such as C.[23]

As an alternative, an application server can run between the Web form and the image management system. With an application server, the first step is to develop Web pages with special tags in the pages. Users can add data, such as a query, to the page and pass it back to the application server. The server replaces the special tags with the results of the calculations or database queries and then sends the completed page to the Web server, which finally sends it to the browser.

A number of these middleware applications are

Computer Network, www.mcn.edu/Standards/mus-ead.html. For MOAC, see *Museums and the Online Archive of California*, www.bampfa.berkeley.edu/moac. For MOA II, see "The Making of America II," sunsite.berkeley.edu/moa2.

23. Special note should be made of recent versions of FileMaker Pro, which include a dynamic Web interface as part of the software. Early reports suggest that the interface does not scale well to large numbers of simultaneous users, but currently it is the easiest Web interface to develop and deploy.

available. Among the most popular are ColdFusion from Allaire and applications that use Microsoft's Active Server Pages (ASP) scripting environment. Recently PHP, a product of the free-software movement, has also been gaining much attention. The three programs are server-side scripting languages, though they respect system resources more than CGI scripts do. All require the system developer to learn a new programming language. For ASP, the task is onerous enough to call for helpful third-party programs, with Drumbeat 2000 the leader in a developing field. If you have programming or Web-development staff willing to learn a new programming language, however, these middleware applications provide the greatest flexibility in interface design.

One final approach to interface design relies on tools that accompany the database. Informix in particular, but also the other large database companies, are actively developing Web-aware plug-ins to their databases. Informix calls their add-ins DataBlades, which allow people to manipulate images by zooming, panning, and so on. Again, the cost in both dollars and technical support to develop and maintain an application in one of these databases is high. On the positive side, datablades let you easily implement many of the functions that users want in a Web interface.

PREPARE FOR CHANGE

Among the numerous possible approaches to an image management system, no solution is right for everyone. While some of these solutions may be right for you, more likely none are perfect. Consequently, you must consider any image management system a temporary solution. The system you adopt today is likely not to be the best solution tomorrow. In addition to maintaining whatever system you do adopt, you must also monitor closely the developments in image database standards in the areas of greatest interest to you. As standards evolve, you must be ready to move your implementation to new platforms and new approaches. Stay consistent—to make data and image migration easier—but remain flexible.

8 Projects to Programs:

DEVELOPING A DIGITAL PRESERVATION POLICY

Oya Y. Rieger

DIGITAL PRESERVATION means retaining digital image collections in a usable and interpretable form for the long term. While "long term" suggests an indefinite future, David Bearman interprets it more usefully as "retention for a period of continuing value."[1] Although "digital preservation" and "digital archiving" are sometimes used interchangeably, here digital archiving refers to initiatives assumed by institutions with a mandated responsibility to maintain digital information for legal, fiscal, evidential, or historical purposes. The goals of digital preservation are:

- *Bit identity*: To ensure content (bit-stream) integrity by:
  - monitoring for corruption to data fixity and authenticity;
  - protecting the content from undocumented alteration;
  - securing the data from unauthorized use, and providing media stability.[2]

- *Technical context*: To maintain interactions among the elements of the wider digital environment by:
  - preserving the context (e.g., ensuring that the metadata files and the scripts that link these files to the image collection are intact);
  - maintaining the integrity of links;[3]
  - monitoring dynamic document creation (e.g., on-the-fly conversion or periodic metadata updates).

- *Provenance*: To maintain a record of the content's origin and history; for example, by describing the source of the digital content and alterations that may have taken place since its creation, such as image and metadata updates or changes to the storage media.

- *References and usability*: To ensure that users (including collection managers) can easily locate, retrieve, and use the digital image collection indefinitely; for example, establishing unique identifiers, such as persistent URLs (PURLs) or using CNRI's Handle System®.

ACCESS VS. PRESERVATION

Some institutions view digitization as a strategy to preserve deteriorating originals. Others digitize primarily to enhance access. Whether the goal is

1. David Bearman, *Archival Methods, Technical Report* 3, no. 1 (Pittsburgh, PA: Archives and Museum Informatics, 1989), p. 17-27. For the concept of "enduring value" for electronic records, see James O'Toole, "On the Idea of Permanence," *The American Archivists* 52, no. 1 (winter 1989): p. 10–25. In a British Library Research and Innovation Centre report, the phrase "long term" is equated with 50 years; John C. Bennett, "A Framework of Data Types and Formats, and Issues Affecting the Long Term Preservation of Digital Material," *eLib: The Electronic Libraries Programme, eLib Supporting Studies*, www.ukoln.ac.uk/services/elib/papers/supporting.

2. For an articulation of the bit-stream concept, see Jeff Rothenberg, "Ensuring the Longevity of Digital Documents," www.clir.org/programs/otheractiv/ensuring.pdf. The authenticity of digital resources is discussed in David Bearman and Jennifer Trant, "Authenticity of Digital Resources: Towards a Statement of Requirements in the Research Process," *D-Lib Magazine* (June 1998), www.dlib.org/dlib/june98/06bearman.html.

3. The typical Web page has 15 hypertext links to other pages or objects and five imbedded objects, such as sounds or images: Peter Lyman and Brewster Kahle, "Archiving Digital Cultural Artifacts: Organizing an Agenda for Action," *D-Lib Magazine* (July/August 1998), www.dlib.org/dlib/july98/07lyman.html.

reformatting for preservation or digitizing for access, institutions are obliged to keep these resources viable. A digital imaging initiative's goal may help set priorities for digital preservation; nevertheless, even an access-oriented collection is not immune to technological obsolescence.

DIGITAL IMAGES AS DIGITAL OBJECTS

Digital objects include both content (digital images) and the accompanying files that give meaning and functionality to it. Metadata files, derivatives in various digital forms, and scripts and programs that determine and activate different behaviors (viewing, printing, searching, etc.) are also an integral part of digital preservation. For example, a digital surrogate of a journal issue may include not only a number of TIFF image files but also metadata recorded in TIFF headers and an external FileMaker® Pro database. As David Levy states, preserving the actual digital object is necessary but not sufficient.[4] It is also essential to preserve those features and behaviors that are deemed crucial.

DIGITAL IMAGE COLLECTIONS AS ARTIFACTS

Digital preservation strategies may inadvertently affect the visual presentation and functionality of digital image collections. The archival community has had an ongoing discussion about the value of electronic records as artifacts and the effectiveness of various preservation strategies in retaining this value. Kenneth Thibodeau, of the National Archives, describes the tension between retaining content and retaining visual and functional characteristics:

Preserving an object means keeping it intact, unchanged. Maintaining a digital object unchanged entails sustaining attributes and methods that bind the object to the technology originally used to create or capture it. Over time this binding will tend to raise greater and greater

barriers to using state of the art technology to access the objects.[5]

For example, when TIFF images migrate to a file format that supports resolution on demand, users can navigate different resolution levels in the multiresolution image, instead of choosing between thumbnail, full-screen, and detail versions. Although the content of the images remains intact, their behavior is slightly different. Such issues as they apply to digital image collections have not yet been fully explored in the digital imaging community.

COMPONENTS OF A DIGITAL PRESERVATION POLICY

One of the main challenges in digital preservation is the pace of technological change. Although such advances promise more effective solutions for our information-management problems, they also set a moving target that hinders long-term planning. We need to assess and manage risks associated with digital preservation based on policies that reflect today's knowledge and technology, but factor in the dynamic and fluid nature of technology. The underlying theme of many preservation principles is the importance of life cycle management, from the initial conception of an imaging project to planning for implementation and sustainability. The key components of a digital preservation policy fall into four categories: organizational infrastructure; policies for selection, conversion, and reselection; preservation strategies; and technology forecasting.

Organizational Infrastructure

Establish digital preservation as an institutional responsibility with committed financial and staff support.
Identify skills and staffing patterns required to implement digital preservation strategies. As emphasized in chapter 9, these strategies should be an integral part of an institutional program with clearly identified responsibilities and timelines. Provide ongoing training and professional development opportunities

---

4. David Levy, "Heroic Measures: Reflections on the Possibility and Purpose of Digital Preservation," *ACM Digital Library, International Conference on Digital Libraries: Proceedings of the Third ACM Conference on Digital Libraries, June 21–23, 1998, Pittsburgh, PA,* www.acm.org/pubs/citations/proceedings/dl/276675/p152-levy.

5. Kenneth Thibodeau, "Resolving the Inherent Tensions in Digital Preservation" (paper presented at NSF Workshop on Data Archival and Information Preservation, March 26–27, 1999), *MU Computer Engineering, Computer Science,* cecssrv1.cecs.missouri.edu/NSFWorkshop/thibpp.html.

Why Is Digital Preservation So Challenging?

Concern about the longevity of digital information has been the topic of many recent popular and scholarly articles, and the challenge and importance of digital preservation has been acknowledged worldwide.[6] The CPA/RLG Task Force on Archiving of Digital Information report helped to spark a number of US, European, and Australian initiatives to identify policy frameworks and research agendas.[7] New groups are forming to explore solutions and to develop and test new preservation strategies. Despite these initiatives, practices that pragmatically address digital preservation are slow in coming.[8]

Technological Obsolescence

A recent survey of the digital preservation needs and requirements of RLG-member institutions ranked technology obsolescence as the greatest threat to digital collections.[9] Many technologies disappear as product lines are replaced and backward compatibility is not always assured. Companies go out of business, orphaning proprietary technologies that support image collections, or major technological advances make is less desirable to maintain older systems. Jeff Rothenberg states that as hardware and software become obsolete, "the digital documents that depend on them become unreadable—held hostage to their own encoding."[10] The vulnerable elements of the technical infrastructure for image collections include:

- Storage media, due to physical deterioration, mishandling, improper storage, and obsolescence[11]

- File formats and compression schemes

- Devices, programs, operating systems, access interfaces, and protocols

- Distributed retrieval and processing tools, such as embedded Java™ scripts and applets

- Gaps in institutional memory due to technical staff turnover

Different Stakeholders, Different Interests

Digital imaging projects involve several stakeholders, including copyright holders, users, funders, collection managers, publishers, and project staff. These parties may have different perspectives and competing concerns.[12] For example, investing in storing and preserving infrequently used resources conflicts with a publisher's objective to maximize revenue.

Organizational and Legal Issues

Another key challenge behind digital preservation is the lack of institutional commitment, and the uncertainty about the distribution of responsibilities. The CPA/RLG report expresses the need for a deep infrastructure for long-term preservation, including certified archives to store, migrate, provide access to digital collections, and to exercise aggressive rescue.

Copyright and fair-use regulations that apply to digital collections are in flux. The legal status of digital collections can change over time, with modifications to copyright law. For example, a library or an archive may have a right to digitize, but no right to refresh, due to copyright restrictions governing the storage media or the necessary retrieval software.

Financial Requirements

Limited resources have always impeded developing and implementing traditional preservation programs. Digital preservation requires substantial computing infrastructure and expertise with unpredictable preservation cycles. Different cost models are emerging, mostly comparing print and electronic preservation. For example, it has been argued that digital images need to be migrated every three to five years, with costs approaching those of the original imaging project.[13] Preservation and access are intricately entwined in the digital realm and, unlike print publications, their separate costs cannot be easily identified. Although centralized archiving is an appealing concept, there are as yet no such initiatives to compare in-house and outsourced preservation models.

6. Michael Day, "Preservation of Electronic Information: A Bibliography," *UKOLN: The UK Office for Library and Information Networking*, homes.ukoln.ac.uk/~lismd/preservation.html.

7. Task Force on Archiving of Digital Information, *Preserving Digital Information: Report of the Task Force on Archiving of Digital Information* (Washington, DC: Commission on Preservation and Access, 1996); also available as "Preserving Digital Information: Final Report and Recommendations" (May 20, 1996), *RLG*, www.rlg.org/ArchTF/index.html.

8. After examining 34 documents to review the state of the art in digital preservation guidelines in a European Commission–funded study, Marc Fresko and Kenneth Tombs concluded that compared to digitization standards, digital preservation issues are underrepresented: Marc Fresko and Kenneth Tombs, "Digital Preservation Guidelines," *DigiCult, Publications*, www.echo.lu/digicult/en/study.html.

9. Margaret Hedstrom and Sheon Montgomery, "Digital Preservation Needs and Requirements in RLG Member Institutions" (December 1998), *RLG*, www.rlg.org/preserv/digpres.html.

10. Jeff Rothenberg, "Avoiding Technological Quicksand: Finding a Viable Technical Foundation for Digital Preservation" (January 1999), *Council on Library and Information Resources (CLIR)*, www.clir.org/pubs/reports/rothenberg/contents.html.

11. There is considerable controversy over the physical lifetimes of media, with figures ranging from 5 to 100 years. Although storage media's life span is an important issue, many now recognize that obsolete equipment for reading media represents a far greater threat than the physical fragility of the digital media themselves.

12. For a discussion of the key stages in the life cycle of a digital resource and how these are influenced by the major stakeholders involved in each stage, see Neil Beagrie and Daniel Greenstein, "A Strategic Policy Framework for Creating and Preserving Digital Collections" (July 14, 1998), *Arts and Humanities Data Service*, ahds.ac.uk/manage/framework.htm.

13. Sue MacTavish, "DoD-NARA Scanned Images Standards Conference," *RLG DigiNews* 3, no. 2 (April 15, 1999), www.rlg.org/preserv/diginews/diginews3-2.html.

for staff in charge of developing and implementing preservation strategies, and budget for it.

If your institution has a preservation department, consider expanding its role to encompass digital collections. For the foreseeable future, digital collections will coexist with print collections. In "Preservation in the Future Tense," stressing the importance of staff training, Abby Smith argues that "a fully staffed preservation department will reflect the hybrid nature of a library's holdings, with staff having, or having access to, the necessary expertise to handle a variety of media."[14]

14. Abby Smith, "Preservation in the Future Tense," *CLIR Issues* 3 (May/June 1998), www.clir.org/pubs/issues/issues03.html.

Understand the importance of a financial commitment to staffing, training, and hardware and software upgrades. This is easier said than done. Long-term preservation of digital information demands significant financial investment. For example, at the Inter-university Consortium for Political and Social Research (ICPSR) the archival operation costs nearly a quarter of a million dollars every year, not including the organizational overhead, with a like amount paid for every major migration of data.[15]

15. Richard Rockwell, "ICPSR Has Long History of Archiving Electronic Information," *Association of Research Libraries, Transforming Libraries* 5: "Issues and Innovations in . . . Preserving Digital Information," www.arl.org/transform/pdi/index.html.

Confronting Obsolescence: An Archival Lifestyle

Janet K. Vavra
Director of Technical Services, Inter-university Consortium for Political and Social Research (ICPSR)

When people began archiving electronic data in the early 1960s, very few foresaw or understood the challenges that would eventually arise in the area of digital preservation. The main steps in the archiving process were creating backup copies of the data, assuring there was adequate documentation, and preparing a description of the data collection (now known as metadata). Few resources were dedicated to maintaining the holdings apart from periodic refreshment done by accessing and reading the files and replacing tapes as warranted. This was considered responsible archiving practice for machine-readable data. At that time, documentation was usually in paper format with copies stored at separate sites.

During the 1970s and early 1980s, technological changes focused more on increasing computing speed and storage capacity than on diversifying formats and software. During the late 1980s, the increasing capability of personal computers offered an array of options that had an impact on how digital archives received and disseminated data. Many software packages now present data in formats that are self-contained or compatible with selected software. This is a very user-friendly approach for those willing to use the bundled software, but it presents problems for preservation and general use. Frequently, data embedded in software can be read successfully only within that package. When archiving such a collection, archivists must decide whether to distribute the data in their original format, thereby requiring users to use the necessary software, or create a more standard file, such as an ASCII file with supporting documentation. Furthermore, in order to store the data in the archive, a decision must be made whether or not to convert the data to a more standard format or leave that decision for later when the software is obsolete.

ICPSR'S Experience
Established in 1962, the Inter-university Consortium for

Political and Social Research (ICPSR) provides access to more than 40,000 unique electronic social science research and instructional files.[1] Beginning in the late 1980s, ICPSR, like its counterparts, found itself in a rapidly changing technological world. The immediate problems were:

- Moving all existing data collections to the emerging technology and assuring they remained accessible and uncorrupted.

- Archiving incoming information that was arriving in ever changing diverse technical formats.

By far the largest task was the rapid migration of nearly 40,000 service files from mainframe media to magnetic disk, which would enable users to download files via the Internet without any interruption in service. Users could now select the specific files they wanted to download and peruse machine-readable documentation online, where available, before downloading or ordering the data. ICPSR also developed in-house software that helped evaluate files after they had been transferred to the new system using a series of benchmarks that would signal if a file might have been corrupted.

Besides access copies (service files) for the user community, ICPSR maintains two copies of each access file and its original source file separately to serve as backups for the archival holdings. This practice is similar to maintaining archival and access copies for digital image collections. These files back up the service files and allow staff access to original files when needed. Since these backup files were also stored on mainframe media, they too had to be transferred to magnetic medium compatible with the new computing environment.

Technical documentation that accompanies data files is essential in the interpretation and use of numeric files and it

CONTINUED ON PAGE 139

While launching a new imaging project, remember to address the sustainability of the collection. Although it is difficult to predict digital preservation expenses, build in funds for it and recognize it as a legitimate and ongoing expense.[16] As articulated by Sarah Thomas in her sidebar to chapter 9, currently most digital imaging projects rely on external funding and are not seen as a part of the day-to-day life of libraries. Most of the funders are interested in immediate access and do not provide for the long-term retention of these digital image collections. Develop strategies that address long-term financial issues even if your funding agency does not support such expenditures.

Seek opportunities for cooperation and collaboration.
Seek out opportunities for coordination and resource sharing at national and international levels. This is another principle that is easier to state than to implement. Although codeveloping digital preservation guidelines and sharing responsibility for preservation is appealing, not many projects exemplify this approach. Institutions should actively seek out partners and be willing to compromise to develop joint ventures that accommodate different needs and interests. In the world of networked and distributed information, the solution to digital preservation is also likely to be distributed.

Consider joining consortial archives for centralized preservation. The RLG/CPA report articulates the virtues of centralized archives with its recommendation for developing certified archives. An example of a corsortial archiving effort is demonstrated by OCLC, which through a contractual agreement with publishers has created an electronic-journals archive, the Electronic Collections Online service, to provide subscribing libraries ongoing access to back issues, even if the publisher terminates its agreement. OCLC intends to migrate the service to new technologies as needed.[17] Another example is the AHDS, which offers a centralized preservation service for its members' electronic resources.[18]

---

16. The American Astrophysical Society provides an inspiring example of factoring in preservation costs: their electronic journal subscription price includes future preservation expenses for periodically refreshing and migrating the files.

17. "How Electronic Collections Online Serves as a Digital Archive," *OCLC*, www.oclc.org/oclc/eco/archive.htm.

18. Arts and Humanities Data Service home page, ahds.ac.uk.

CONTINUED FROM PAGE 138

must remain accessible and usable. In 1997, ICPSR received preliminary funding to preserve paper copies of its documentation as well as to provide a standard for future Web-based documents. Known as the Data Documentation Initiative (DDI), the project's aim is to develop an XML-based markup standard to allow more meaningful searching of electronic documents on the Web.[2] Among the goals of this project is assuring the preservation of documentation and metadata while facilitating online searching of both content and logical relationships among elements. This project also involves the conversion of paper copies of the existing documentation into PDF format to provide access copies for users.

Conclusions

▸ Preservation ultimately means a financial commitment to staffing, training, hardware, and software. Archives should routinely build long-term preservation into their archival workflow and budgets.

▸ We may not be able to preserve everything. Although it may be a challenge to determine which collections warrant preservation efforts and which do not, it is imperative that we make the effort to identify digital holdings that are essential to preserve. Archives should work with committees of users and experts in setting priorities for this effort.

▸ Preserving technology is not a viable digital preservation strategy for data archives. We may think it is costly to migrate and translate information now, but it is even more expensive and difficult to maintain the technology that many valuable collections now need to be viable. The idea of maintaining hardware and software over time to be able to read legacy files is impractical. Just the space needed to store all the hardware that has existed over time is overwhelming, not to mention the personnel needed to understand and operate it.

▸ It is important not to delay technological migration until the technology being replaced is several generations old. Generally, vendors provide for relatively simple migration from an earlier version of software and hardware, but the window of opportunity closes quite quickly after the introduction of the replacement technology.

▸ Data archives, such as ICPSR, have strong commitments to both their depositors and their users to make holdings available indefinitely. In our litigation-prone society, legal considerations will certainly enter into future decisions about preservation and migration of materials.

Notes
1. ICPSR home page, www.icpsr.umich.edu.
2. The DDI project ("The Data Documentation Initiative: A Project to Develop an XML Document Type Definition for Data Documentation," *ICPSR*, www.icpsr.umich.edu/DDI/codebook.html) is supported by funding from the National Science Foundation Grant SBR-9617813DDI.

Principles for Creating a Basic Preservation Strategy

Margaret Hedstrom
Associate Professor, School of Information, University of Michigan

Librarians, archivists, curators, and others who are concerned with providing continuing access to information are seeking more reliable and cost-effective methods for digital preservation. Even when improving access is the main goal, effective digital preservation strategies are needed to protect investments in digitization and provide ongoing access.

Focus Current Efforts on Risk Reduction

Organizations can reduce the risk of losing digital materials to accidents, disasters, neglect, and technology obsolescence. While with our current knowledge and experience we cannot guarantee the durability of digital materials far into the future, the immediate goal of digital preservation is risk reduction while we seek more lasting solutions. Today's solutions will not necessarily be the best solutions in the future, but waiting for ideal solutions will place much of the digital information we are generating at great risk. Every digital imaging program should include a plan for preserving the digital materials it creates.

There is no consensus about the most effective digital preservation strategy although some methodologies work effectively for certain classes of digital materials. Given the wide variety of digital formats and ongoing change in capture, storage, and management technologies, no single approach to digital preservation is likely to satisfy all requirements for all types of digital materials. Therefore, a consensus on basic digital preservation strategies does not necessarily imply agreement on the one best way to preserve digital materials. Rather, it entails agreement on a manageable set of interim solutions that accommodate the varying requirements for different digital formats.

Digital preservation strategies such as migration, emulation, and digital archeology are being discussed and investigated. Each of these approaches has a potential role in digital preservation, but does not guarantee longevity. Rather than searching for the one best method for permanent preservation of digital materials, we should direct our efforts toward refining existing methodologies, developing and testing alternatives or complementary solutions, and reaching agreement on how they can work together across the information life cycle.

Safeguard Digital Masters

Archival master files should contain the highest-quality scan you can afford, stored in a format that conforms to adopted standards with no compression or with lossless compression. The safest course is to select storage formats that conform to de facto standards and to use products that are widely deployed in the market. Keep in mind that different standards apply to different aspects of digital information, such as encoding, tonal and color representation, structuring and organizing data, interfaces, and storage media. Products that conform to standards in each of these areas tend to be more durable, but products and standards in all of these areas are evolving rapidly.

Develop clear specifications and statements of responsibility for maintenance of the archival files of images and metadata. Storage might be handled by the organization that generated the files, by an institution's information technology services unit, or by a third party. The critical issue here is that the responsibilities are clearly defined and that specifications are established for storage conditions, backup, recovery, and long-term maintenance. Online, near-line, and offline storage are all acceptable for maintaining the archival master files. The best solution will depend on how the master file is used, how frequently it is updated, the types of storage services available, and the cost.

Maintain Metadata Separately from the Master Image Files

Specifications for storage formats should cover the digital images and any associated administrative, structural, and descriptive metadata. Maintaining metadata separately from the master files of digital images will simplify long-term preservation. The products used to manage metadata and those used for the digital image files may evolve at different rates. With just enough overlapping data to link the metadata to the images, image files and metadata can be updated or migrated independently. To manage metadata, most organizations use commercially available database or word processing packages that are customized for the requirements of a specific project or institution. Select word processing packages and database systems that can save files in several different formats or export data in common interchange formats, such as ASCII delimited files.

Recognize That Loss Is Inevitable

Copying supports long-term preservation when deterioration of the physical medium threatens the longevity of the data, but the data formats remain stable. Sometimes, the files must be migrated to a new format or transferred to a new medium that can be read and interpreted by readily available systems. Such conversions may entail loss of information, loss of functionality, or other changes in the file. If conversions are necessary, select new formats that minimize these losses and carefully document the steps in the conversion process.

Retain Originals or Create Analog Backups

Most digital materials that are created through reformatting serve as surrogates for the original documents. Until digital preservation strategies are perfected and we have more experience in long-term maintenance, it is prudent to retain originals or analog surrogates so that they could be rescanned in the event of a catastrophic loss of the digital files. If migration to a new format would involve unacceptable losses or threaten the integrity of the archival master file, rescanning may be the only acceptable alternative.

Rely on each other's expertise and experience. Many institutions are dealing with similar preservation challenges and should be sharing strategies such as migration paths developed to upgrade systems.

Policies for Selection, Conversion, and Reselection

Select and create digital image collections with long-lasting value.

▸ Address long-term preservation concerns at the onset, starting with selection. Select collections for digitization considering the issues involved in their long-term maintenance.

▸ Create image collections that meet baseline standards for quality and functionality. The long-term viability of digital image collections should be based on content and usability, not limited by technical decisions made at any point in the digitization process.[19] It may not be worth preserving poor-quality digital images that barely support user needs.

▸ Develop or adopt guidelines and standards to be used throughout your digitization programs, including conversion, metadata creation, processing, and quality control. Document these procedures and apply them consistently. It is much easier to develop preservation strategies for uniform, consistent digital collections. While digital imaging and preservation standards and guidelines are slow in coming and there are few national or international standards to follow, set your own and apply them consistently throughout your organization.[20] For example, with funding from the Institute of Museum and Library Services (IMLS), Cornell University Library is developing metadata standards for its digital image collections. The National Library of Australia is also actively involved in devel-

oping a set of preservation metadata elements. These interim procedures will create a level of control over your institution's digital preservation procedures, facilitating migration to upcoming national or international standards.

▸ Use sparingly the techniques and methods created to solve today's problems. For example, both compression and encryption schemes add an extra level of complexity and may not work well in future information environments. If lossy compression is desired, it is better to adopt a standard compression format and set the compression ratio low to attain high-quality images.

▸ Avoid shortcutting certain steps during the initial digitization stage to control expenses and beat the project deadline. Creating images with long-lasting value will prove to be cost-effective in the long run. For example, use of color bars during image capture may help rebalance color in future viewing environments.

▸ Keep the master copies at the heart of your preservation strategy. As described in chapter 6, relying on on-the-fly conversion to create access copies, instead of creating static derivatives just in case they are needed, proves to be prudent in the long run. Although it is ideal to create image collections that meet both archival and access needs, sometimes this may be unattainable. For example, ICPSR separated archival and access copies of data sets in order to keep master files stabilized and still address changing usage patterns.

▸ Use standard and system-independent file formats if possible. There is no universal standard format for digital images similar to ASCII for raw text or SGML for encoded text, so rely on formats that are widely used and supported (see table 8.1). Although SGML, especially within the framework of the Text Encoding Initiative, has successfully captured content using a standard format, it does not accommodate digital images with pictorial content, illustrations, or significant formatting information.[21] If standards are lacking, rely on common practices.

19. Cornell advocates full-informational capture to ensure that digital objects are rich enough to continue to be useful; see Stephen Chapman and Anne R Kenney, "Digital Conversion of Research Library Materials: A Case for Full Informational Capture," *D-Lib Magazine* (October 1996), www.dlib.org/dlib/october96/cornell/10chapman.html.

20. Jeff Rothenberg argues that "although defining ultimate standards for digital documents is an admirable goal, it is premature as the information technology is still on the steepest slope of its learning curve": Rothenberg, "Ensuring Longevity."

21. For an exploration of the suitability of SGML for access to digital libraries, with special emphasis on preservation issues, see James Coleman and Don Willis, *SGML as a Framework for Digital Preservation and Access* (Washington, DC: Commission on Preservation and Access, 1997).

Table 8.1. Characteristics of File Formats for Long-Term Utility

General Recommendations:
- Thorough, nonproprietary development and documentation
- Proven backward compatibility and reliability
- Wide adoption by large consortia and groups, to increase the chances for well-defined migration paths
- Support for exchange standards such as the Electronic Document Interchange Standard

Technical Requirements (Image Quality):
- Support for bit-depth greater than 24 for continuous-tone documents
- Various lossless and lossy compression schemes
- Support for various color models and color management features, such as gamma/white point and ICC profiling)
- Multiresolution capability to support both master and access copies

Metadata:
- Features to encompass/encapsulate various kinds of metadata
- Support for relationship among various components (images, metadata files, scripts, etc.)
- A flexible architecture for metadata recording (e.g., file headers, links to external files, etc.)

Hardware/Software:
- Platform-independent viewing/retrieval software (or cross-platform consistency)
- Minimal hardware/software dependencies
- Abundant retrieval and image-processing programs for several platforms.

Security:
- Features to report data corruption or error detection and correction
- Features for authentication

For example, many recommend saving images in the most common file formats (e.g., TIFF), since they predict the future availability of file converters or emulators for commonly used file formats.

Set priorities.

Develop reselection criteria to identify digital image collections for long-term retention, similar to those created for preserving print collections. A recurring theme in digital preservation literature has been that we cannot save everything. Not every digital collection warrants long-term preservation, and some options may be costly or ineffective.[22] Different models are emerging for selection for preservation. Margit Dementi states that increased access is no longer an enemy of preservation as high-use items are the most likely to be preserved.[23] The demand for them will ensure that they are migrated and will reduce the risk of their being lost. However, John Price-Wilkin, in chapter 6, supports the claim that only a small percentage of images in a digital image collection are used. The challenge is how to ensure that less-used materials with significant scholarly value are preserved. These questions help to assess the value of a digital image collection:

- Is there a physical counterpart (e.g., print, microfilm) to the digital collection?

- Can the collection be digitized again (what is the cost-benefit analysis of redigitization and preservation)?

- Does the collection fall within one of your institution's subject strengths?

- Do other institutions hold similar digital copies? If yes, what does the collection at hand contribute?

- Does the collection complement other collections housed elsewhere?

- What behaviors and functionalities of the collection are essential to retain? Is it possible to retain them through preservation?

- Is the collection self-sustaining or income generating?

- How often is the collection being used? How does the usage pattern compare to other collections?

- Technically, is the collection sound enough to warrant preservation? How good are the images and metadata?

22. For example, the National Library of Australia states that "in an ideal world, deposit libraries might preserve everything, but operating in an environment of technical and resource constraints means that priorities must be set": "A Draft Research Agenda for the Preservation of Physical Format Digital Publications," *National Library of Australia*, www.nla.gov.au/policy/rsagenda.html.

23. Margit R. Dementi, "Access and Archiving as a New Paradigm," *The Journal of Electronic Publishing* 3, iss. 3 (March 1998), www.press.umich.edu/jep/03-03/dementi.html.

- What are the consequences of losing this collection?

- Is there any legal or consortial obligation to maintain this collection?

- Are the purposes that drove the initial selection for digitization still valid, or are there any new ones?

Create metadata to support future preservation strategies.

As elaborated in chapter 5, capture and record metadata that describes the technical attributes of image collections and documents the policies and events in their life cycles.[24] Among the goals of preservation metadata are:

- Providing a persistent and unique ID, using, for example, the Handle System, a PURL, a Uniform Resource Name (URN), or the Digital Object Identifier (DOI) system.[25]

- Ensuring the integrity of image files; for example, recording the original checksum value.[26]

- Facilitating image quality assessment to support retention decisions.

- Maintaining provenance data; for example, the Internet Engineering Task Force's Uniform Resource Characteristics (URCs) proposes to include intellectual property rights, provenance, and context information.

- Providing reference information to facilitate the identification and utilization of the resource; for example, structuring labels, descriptive data.

- Identifying intellectual property rights.

- Verifying the authenticity and integrity of digital objects.

- Providing information about hardware and software to determine dependencies.

The Digital Library Federation–sponsored report *Preserving the Whole*, which focused on data files, provides an excellent example of the value of metadata in preservation. Column binary format used for numeric files, although technically obsolete, was so well documented that the project staff could not only migrate these data sets into other formats, but also read them in their native format.[27]

A key issue is not only what to record, but how and where to record it to facilitate maintaining and updating metadata. The approval of the MARBI Proposal No. 99-01 allows recording of preservation and reformatting information in the MARC 007 field for computer files. This information enhances the retrieval and management of digitally reformatted materials and guides decisions to digitize materials for preservation purposes.[28]

Safeguard metadata that will support future preservation activities. Do not rely on institutional memory for information that is essential for preservation. Some metadata files have dynamic content that is continually corrected or updated as information is added about refreshing and migration cycles, information loss, or upgrades to the access system. This information should be systematically recorded, updated, and secured.

Preservation Actions

Store digital media with care.

House images and accompanying files in secure, reliable media and locations.[29] Store and handle these files according to guidelines that optimize their life expectancy. Consider storing master files

24. Michael Day provides a summary of the current preservation metadata issues and initiatives in Michael Day, "Issues and Approaches to Preservation Metadata" (paper presented at the Joint RLG and NPO Preservation Conference, Guidelines for Digital Imaging, September 28–30, 1998), *RLG*, www.rlg.org/preserv/joint/day.html.

25. The Handle System® home page, www.handle.net; Persistent URL home page, purl.oclc.org; "rfc2276: Architectural Principles of Uniform Resource Name Resolution," (January 1998) *The Ohio State University, Computer and Information Science*, www.cis.ohio-state.edu/htbin/rfc/rfc2276.html; DOI: The Digital Object Identifier home page, www.doi.org.

26. A checksum is commonly used to track the bit-level integrity of digital files and ensure that they are identical to the original.

27. Ann Green, JoAnn Dionne, and Martin Dennis, *Preserving the Whole: A Two-Track Approach to Rescuing Social Science Data and Metadata* (Washington, DC: Council on Library and Information Resources, 1999), www.clir.org/pubs/reports/pub83/contents.html.

28. "FAQs" and "RLG News," *RLG DigiNews* 3, no. 1 (February 15, 1999), www.rlg.org/preserv/diginews/diginews3-1.html.

29. The National Media Laboratory no longer publishes charts with comparative media longevity due to the challenge of keeping up with the frequent media improvements. Many storage media manufacturers and organizations now publish media longevity information on their Web sites; for example, the SIGCAT (Special Interest Group on CD/DVD Applications & Technology) Foundation home page, www.sigcat.org. SIGCAT is a user group dedicated to educating the public and promoting the growth of applications based on CD and DVD technologies.

offline or creating backups on separate servers. Periodic and systematic reading and backup is essential. Your system operators may be able to set parameters for your system to execute these procedures automatically. The British Library Research and Innovation Centre suggests storing archival copies on industry standard digital tape or on other approved reliable media as they emerge.[30] The report also recommends:

▸ Storing offsite copies far enough from onsite copies that they are immune from any disaster affecting the onsite copies.

▸ Writing several archival copies with different software to protect them against corruption from malfunctions, viruses, or bugs.

▸ Checking media periodically for their readability.

▸ Checking the integrity of data files periodically, using checksum and similar procedures.

Create reliable backups for your digital collection. Duplication is a simple way to avoid losing files to deterioration or corruption. Copy the digital files onto newer, fresher media before the old media becomes vulnerable due to aging or storage conditions.

Adopt incremental, ongoing maintenance, which may be the most cost-effective strategy for the long run. Upgrade the system components and file formats incrementally to maintain a robust system.

Evaluate and implement preservation strategies.
Select preservation strategies based on factors such as image attributes, cost-effectiveness, user and access requirements, effective preservation of the essential image features, staffing, and hardware/software environments. Currently, no correct or ideal preservation strategy accommodates all digital image collections. See page 146 for the pros and cons of five different approaches.

Understand that current digital preservation strategies are merely tools to control risks. For instance, table 8.2 lists the risks involved in file-format migration. To be able to use such a risk-assessment matrix effectively, you need to forecast the impact of these factors (e.g., catastrophic, critical,

marginal, and negligible) and to identify their probability.

Consider a hybrid approach.
Some argue that none of the current digital preservation strategies are reliable enough, therefore the only safe measure is to retain the original or create analog backups, for example, printouts from images or microfilm.[31] These hybrid measures are seen as necessary until there is a clearer understanding of the processes, costs, and risks involved in digital preservation. Microfilm is a well-established preservation medium. For example, computer output microfilm (COM) can serve as an analog safeguard from which the digital files can be recreated by scanning.[32] A recent working paper argues that microfilming remains the only viable long-term strategy for dealing with the preservation problems posed by brittle paper.[33] The analog strategy was deemed legally sufficient with the approval of NARA's General Records Schedule 20 (GRS 20), which authorizes federal agencies to dispose of certain electronic files after copying them to a paper, microform, or electronic record-keeping system.[34]

Technology Forecasting

Assess risks by monitoring technological changes.
Keep an eye on technological changes, especially those with potential impact on digital preservation. Currently, digital preservation is a manual, labor-intensive process. There are not yet any systems incorporated into digital library architectures or

30. Beagrie and Greenstein, "A Strategic Policy Framework."

31. Norsam Technology's HD-ROSETTA is another analog backup option with a technology to write micro-images on silicon wafers: Norsam Technologies home page, www.norsam.com.

32. Anne R. Kenney, *Digital to Microfilm Conversion: A Demonstration Project, 1994–1996, Final Report* (Ithaca, NY: Cornell University Library, 1997); available from "Publications," *Cornell University Library, Preservation & Conservation*, www.library.cornell.edu/preservation/pub.htm.

33. Stephen Chapman, Paul Conway, and Anne R. Kenney, "Digital Imaging and Preservation Microfilm: The Future of the Hybrid Approach for the Preservation of Brittle Books," *CLIR, Commission on Preservation & Access*, www.clir.org/cpa/archives/hybridintro.html, examines the dual use of microfilm for preservation and digital imaging for enhanced access in the context of the brittle books program, comparing film-first and scan-first strategies.

34. National Archives and Records Administration, Electronic Records Information/GRS20, "Electronic Records Work Group," *NARA*, www.nara.gov/records/grs20/index.html.

Table 8.2. Risks Associated with File Format–Based Migration[35]

Risk	Examples
Content fixity—*bit configuration, including bitstream, form, and structure*	Bits/bitstreams are corrupted by software bugs or mishandling of storage media, or mechanical failure of devices.
	New compression alters the bit configuration.
	File-header information does not migrate, or migrates partially or incorrectly.
	Image quality (resolution, dynamic range, color spaces, etc.) is affected by alterations to the bit configuration.
	New file format specifications change the byte order.
Security	The watermark, digital stamp, or other cryptographic techniques are affected.
Context and integrity—*relationship/interaction with other files or other elements of the wider digital environment, including hardware/software dependencies*	Different hardware and software are required to read and process the new file format.
	Links to other files (metadata files, scripts, derivatives such as marked-up or text versions, on-the-fly conversion programs) are altered.
	Reduced file size, due to file-format organization or new compression, causes denser storage, with potential directory-structuring problems if you try to consolidate files to use the extra storage space.
	The medium becomes denser, affecting labels and file structuring; this might be also due to file-organization protocols of the new storage medium or operating system.
References—*location of images among other digital objects*	File extensions change.
	Migration activity is not well documented, leading to incomplete or inaccurate provenance information and potential problems for future migration.
Cost	Long-term costs are unpredictable since each migration may differ fundamentally in nature.
	The value of the collection may not be sufficiently determined, making prioritization impossible.
	Models are unscalable without a standard architecture (centralized storage, metadata standards, file format/compression standards, etc.) that allows the same migration strategy to be easily implemented for other similar collections.
Staffing	Staff turnover affects the continuity of migration decisions and long-term planning, especially with insufficient metadata and a poorly documented migration path.
	Institutional support may waver between making digital preservation a full-time, permanent job responsibility or an ad hoc, temporary assignment for rescue operations.
	Staff may lack technical expertise.
	The unpredictability of migration cycles makes planning for staffing requirements (skills, time, money, etc.) a challenge.
Functionality	Features introduced by the new file format may affect the creation of derivatives, such as printouts.
	If the master copy is also used for access, changes may decrease or increase functionality, requiring modifications to the interface.
	The new format may not support unique features, such as the progressive display functionality of GIF files.
	The value of the original as an artifact may be lost.
Legal	Strict copyright regulations may limit the use of new derivatives that can be created from the new format.

35. The author developed this matrix as a part of a CLIR-funded project at Cornell, Risk Management of Digital Information. The deliverables of the study, including a migration-risk assessment tool for file-format-based migration will be available from "CLIR Publications," *CLIR*, www.clir.org/pubs/pubs.html.

image databases that automatically examine the well-being of digital collections and signal potential problems, except for some simple processes such as automated data corruption reporting. Cornell's PRISM and Stanford's Archival Digital Library Architecture projects promise to integrate some preservation functions within a digital library architecture. Even when there are automated systems, technology scouting will always involve human monitoring and judgment, such as investigating upcoming file formats like JPEG2000, TIFF 7.0.

Assess when to implement preservation strategies: neither too soon nor too late. The window of opportunity may close quickly after the replacement of the technology. For example, the Cornell University Library has a group of image files in TIFF 5.0 format. Comparison of the TIFF 5.0 and 6.0 file structures revealed no current risks in not upgrading the images, so they have not been migrated.

DIGITAL PRESERVATION STRATEGIES

Refreshing

Refreshing involves copying content from one storage medium to another. The Open Archival Information System (OAIS) defines refreshing more narrowly as copying all bits exactly to an identical medium.[36] Refreshing targets only media obsolescence and is not a full-service preservation strategy. It should be seen as an integral part of ongoing maintenance for digital collections.

Why Refresh?

▸ As preventive care, to control aging and decay that make storage media unreliable and vulnerable.

▸ To avoid media obsolescence.

▸ To amalgamate collections for better integration and management.

▸ To take advantage of more efficient and more reliable storage technologies that offer higher capacity, better access, and lower price.

Challenges

▸ The unpredictable schedule for refreshing, which depends on several factors such as the frequency of use and storage conditions, hinders long-term planning.

▸ Refreshing may introduce copyright issues if the usage is limited to a given storage medium for access control.

▸ Refreshing to a new medium may unintentionally change bits.

▸ As media become dense, it is tricky to maintain any labeling information that is associated with the original physical media. For example, content in several CD-ROMs will fit in one DVD, so the merged content needs to be relabeled correctly.

Migration

Migration is the process of transferring digital information from one hardware and software setting to another or from one computer generation to subsequent generations. Charles Dollar, referring to electronic records, makes a distinction between conversion and migration. Conversion involves exporting digital information from one software environment to another with little or no structural loss, and no content or context loss (e.g., system software upgrades or reformatting a text-based file in a standard format such as SGML). Migration, on the other hand, involves more risks since neither backward compatibility nor export/import gateways exist between the original system and the new one.[37]

Why Migrate?

▸ To increase system functionality as the technology evolves; for example, adopting an on-the-the fly conversion program may require the

36. Don Sawyer and Lou Reich, *Reference Model for an Open Archival Information System (OAIS)* CCSDS 650.0-W-5.0, White Book, Issue 5, April 21, 1999; see "ISO Archiving Standards— Reference Model Papers," ssdoo.gsfc.nasa.gov/nost/isoas/ref_ model.html.

37. Charles M. Dollar, *Authentic Electronic Records: Strategies for Long-Term Access* (Chicago, IL: Cohasset Associates, Inc., 1999), pp. 29–33.

migration of image files into a new file format or switching to a new storage system that speeds up delivery.

- To rescue digital data that are on the verge of obsolescence.

- To upgrade the system that has a known problem or defect, such as software bugs.

- To create a more stable, reliable technical environment.

- To increase the functionality of the collection and possibly save time and money; for example, using a file format that is supported by Web browsers.

- To consolidate collections in the same technical infrastructure.

Challenges

- Like refreshing, migration is difficult to plan for. It requires constant vigilance on the part of the staff to monitor and predict changes that will threaten a digital collection's security, such as changes to file formats, retrieval software, operating system, and hardware.

- Migration is labor intensive, requiring thorough planning and analysis, and it does not scale well. Each cycle may be a new experience,

---

Refreshing: Lessons Learned in a Library
Deborah Woodyard
Digital Preservation Officer, National Library of Australia

Periodically transferring images to new media, or refreshing, is a recommended preservation strategy for maintaining the longevity of digital image collections. When the National Library of Australia (NLA) undertook a media-transfer project to understand the processes and risks associated with refreshing, the transfer process proved to be much more complicated than expected. Although this study focused on transfer from floppy disks to CD-R, some of the findings can be generalized to refreshing projects that involve other electronic media.

The NLA receives hundreds of publications every year that contain CD-ROMs and floppy disks. Media instability and technological obsolescence threaten to greatly reduce the life of these publications. CD-ROMs are still current, popular, and relatively stable; however, the average life of a floppy disk is estimated at less than ten years, so the NLA has initially focused attention on this storage format. For a more stable carrier, the NLA chose CD-R because it is common and easy to use, with better life expectancy than magnetic media. Copyright restrictions make it essential to use a physical format carrier rather than online storage. Online storage could imply possible networking of the material, which would contravene copyright.

Before embarking on the refreshing project, the NLA designed a method for what initially appeared to be a straightforward task: identifying and locating disks in the library's collections, selecting a representative sample, testing whether the disks were still operational, copying the disks in the sample to CD-R, and testing whether the data were still functional. At many points in the process, however, unforeseen issues proved this method was more complicated than we had anticipated. The first challenge was to find the floppy disks in the collection. We searched the catalog using the general material designator (GMD) "computer file," and keywords such as "disk" and "disc"

in the collation fields. It became apparent that we need more detailed information than is currently available in catalog records about the type of disks and their system requirements.

Results

Fortunately, we did not face problems we might have expected, such as computer viruses and technical difficulties in copying. However, technological obsolescence was a major challenge and many disks could not be used because appropriate hardware or software was no longer available. This meant that a significant proportion (36%) of sampled items could not be copied and 19% of the copied items could not be tested to see if they were readable both before and after copying. Changing technology also meant that some items (6%) had a particular path hard-coded, such as the A: drive or root directory, and would not work from the address of the copied version.

Although affecting only a very small number of items, these problems were also encountered:

- Some disks were unusable because of deterioration most likely caused by exposure to magnetic fields or natural media degradation.

- One publication had blank disks.

- Some data had no intuitively obvious way to be used nor any available information.

- A security tag had been stuck over the metal sliding cover of a 3½" disk, preventing its use.

Out of the 64 items:

CONTINUED ON PAGE 148

CONTINUED FROM PAGE 147

▶ 24 were found to be nonfunctional before transfer was even attempted.

▶ 5 were nonfunctional after they had been transferred. They were copied to the CD-R using incompatible software for their system requirements, so their data were inadvertently scrambled.

▶ 13 were not supported by currently owned software and therefore could not be run before or after transfer. However, file comparison after transfer suggests the copying was faithful. If the supporting software were acquired by the library, these items should still be readable.

▶ 22 were fully functional before and after transfer to the new medium.

The NLA now has more informed views of media transfer, including the variety of hardware and software required by the collection and the specifics of the copying processes. This knowledge has been instrumental in implementing media transfer as a standard method in preserving and managing electronic resources stored on floppy disks. Currently, a CD-R backup or "preservation surrogate" is made for all the newly acquired floppy disks and this process has also been applied retrospectively where possible. An ASCII text file is created to accompany each publication backup. This text file contains a description of the contents of the files, the systems requirement details, a record of the original format, and any instructions on how to use this material following the change in storage format. The text file is saved with the publication on the new medium and in the future will also be recorded in a searchable database. The disks are packaged with a notice to redirect use to the preservation surrogate and their catalog records include a note about where the surrogate can be found. Catalog notes can be changed in batch mode when data are transferred to another new medium or location.

This project uncovered the importance of recording and maintaining technical information about the software and hardware requirements of digital publications. This preservation and administrative metadata needs to be recorded in a searchable format so that one can identify all the digital resources that are recorded on a particular vulnerable medium or that rely on software and hardware technologies that are becoming obsolete. Other lessons include ensuring that appropriate software is available for the copying procedure and maintaining the essential functions of data related to the current format, such as hard-coded drive names or fixed data locations, as will be the case with CD-ROMs.

The NLA staff will continue to monitor the changes in storage technology, and digital content will be transferred to more reliable storage media as they become available. Perhaps the most important issue this trial highlighted was that refreshing strategies involve unexpected complications and therefore do not guarantee preservation, but they are a necessary step to minimize risks. Even media transfer, which is considered less complicated than migration, involves hidden risks.

making long-term preservation planning difficult. Scalable automated or semiautomated migration techniques are rarely possible.[38]

▶ Efforts and risks involved will vary depending on the complexity of the process. Transferring a group of TIFF 4.0 images to TIFF 5.0 may be easy, but converting a group of metadata files from an obsolete proprietary format to a new format may be demanding.

▶ Missing the window of opportunity makes it infeasible to migrate the files.

▶ Successive conversions may corrupt a digital document or substantially affect how it was experienced by its author and original audience.[39]

▶ Sometimes information loss is inevitable.[40] Subtle losses may be difficult to detect and assess. This is much more serious than migration affecting the behavior of images.

▶ The absence of unambiguous specifications of the source and target environment (often due to protection of proprietary interests) significantly contributes to migration's high cost and low reliability.[41]

▶ There may be legal restrictions on migration, such as prohibitions against reformatting the digital files into a different format.

38. John Mark Ockerbloom's Typed Object Model is an automated tool that facilitates migration; it includes agent-based tools for migrating digital materials from obsolete platforms to new ones, supporting several data formats: "John Mark Ockerbloom's Thesis Page," www.cs.cmu.edu/~spok/thesis.html. Ockerbloom recently joined the University of Pennsylvania and intends to develop an online conversion service to support migration of data files in digital archives (John Mark Ockerbloom, e-mail to Oya Y. Rieger, September 10, 1999).

39. Jeff Rothenberg, "Using Emulation to Preserve Digital Information" (paper presented at NSF Workshop on Data Archival and Information Preservation, March 26–27, 1999), *MU Computer Engineering, Computer Science*, cecssrv1.cecs.missouri.edu/NSFWorkshop/ppaper3.html.

40. For example: "Recently the Food and Drug Administration said that some pharmaceutical companies were finding errors as they transferred drug-testing data from Unix to Windows NT operating systems. In some instances, the errors resulted in blood-pressure numbers that were randomly off by up to eight digits." Arlyn Tobias Gajilan, "History: We're Losing It," *Newsweek* (July 12, 1999), www.newsweek.com/nw-srv/printed/us/st/ty0102_1.htm.

41. David Bearman, "Reality and Chimeras in the Preservation of Electronic Records," *D-Lib Magazine* 5, no. 4 (April 1999), www.dlib.org/dlib/april99/bearman/04bearman.html.

Emulation

Emulation involves the re-creation of the technical environment required to view and use a digital collection.[42] This is achieved by maintaining information about the hardware and software requirements so that the system can be reengineered. The original data are untouched, except for periodic media refreshment. Jeff Rothenberg argues that the only way to ensure the appropriate behavior of a digital document is to run the original software used to view the document in its native form.[43]

Why Emulate?

- To support the original intent of a digital document by recreating its original functionality, look, and feel.

- To develop a scalable and extensible preservation strategy for other similar collections.

- To have a backup strategy for migration.

Challenges

- Emulation may be technically infeasible and requires a leap of faith regarding future technologies and staff skills.

- It may be very costly, with no guarantees.

- It presumes that users will understand how to use old systems.

- It raises potential intellectual-property issues about acquiring and maintaining proprietary information necessary to emulate hardware and software.

- Troubleshooting for software bugs and viruses is very difficult.

- Emulation may focus on preserving functionality rather than the integrity and authenticity of data.[44]

- Success depends on creation and capture of metadata.

- Some hardware components, such as RAM, video and sound cards, and even analog converters are difficult to emulate.

- Media refreshing is a part of the emulation strategy, with its own risks.

Technology Preservation

Technology preservation involves preserving the technical environment that runs the system, including software and hardware such as operating systems, original application software, media drives, etc.[45] It preserves access tools and involves refreshing the content to new media as needed. While technology preservation means preserving the technical environment rather than reengineering it, as emulation does, many of the same issues apply to both.

Why Preserve Technology?

- So that it can be used as "a relatively desperate measure in cases where valuable digital resources cannot be converted into hardware and/or software independent formats and migrated forwards."[46]

- Like emulation, to preserve the digital objects' original behaviors and form.

Challenges

- In the long term, technology preservation is likely to be expensive and impractical, because of space requirements, maintenance costs (software bugs, trouble shooting, replacement of

---

42. For detailed information on various aspects of emulation, including its advantages and disadvantages, see Rothenberg, "Avoiding Technological Quicksand"; Rothenberg, "Ensuring Longevity"; Rothenberg, "Using Emulation"; Thibodeau, "Resolving the Inherent Tensions."

43. The Universal Preservation Format (UPF) project takes a related approach, aiming to prototype a hardware- and software-independent format to ensure the accessibility of a wide array of data types—especially video formats—into the indefinite future. One of the goals of the proposed UPF metadata is to store all the technical specifications required to build and rebuild media browsers to access contained materials throughout time; the UPF home page, info.wgbh.org/upf.

44. Bearman, "Reality and Chimeras."

45. Illustrating this approach, the Computer Conservation Society in the UK aims to promote the conservation of historic computers: Computer Conservation Society home page, www.cs.man.ac.uk/CCS.

46. Tony Hendley, *Comparison of Methods & Costs of Digital Preservation: A Consultancy Study*, British Library Research and Innovation Report 106 (London: British Library and Innovation Centre, 1998).

aging hardware components), the natural aging of computers, and the maintenance of staff skills.

▸ It is not scalable and cannot be used for all digital collections.

Digital Archaeology

Digital archaeology involves methods and procedures to rescue content from damaged media or from obsolete or damaged hardware and software environments.[47]

Why Digital Archaeology?

▸ To rescue content after missed refreshment and migration opportunities, or from damaged media or technical infrastructures that are unknown or no longer available.[48]

Challenges

▸ Since there are no documented examples of this technique, its success rate and resource requirements are unknown.

▸ It is not scalable and could be very expensive.

PROMISING DIGITAL PRESERVATION INITIATIVES

Basic Image Interchange Format (BIIF)

164.214.2.51/ntb/baseline/docs/biif/index.html
BIIF is an ISO/IEC (the International Electrotechnical Commission) standard in development to provide a foundation for interoperability in the interchange of images and image-related data. The Archiving and Distribution Working Group works on the development of a BIIF profile to demon-

strate the capabilities and potential of BIIF for the library and archival preservation communities.[49]

Cedars (CURL Exemplars in Digital Archives) 1998–2000

www.curl.ac.uk/cedarsinfo.shtml
Part of the UK Electronic Libraries Programme (eLib) and managed by CURL (Consortium of University Research Libraries), the Cedars Project aims to produce strategic frameworks for digital collection management policies, and to promote methods appropriate for long-term preservation of different classes of digital resources, including the creation of appropriate metadata. Cedars's distributed archive architecture is based on an implementation of the OAIS reference model.

Cedars 2: Emulation Options for Digital Preservation, 1999–2001

www.leeds.ac.uk/cedars/cedars2/index.htm
Funded by the Joint NSF/JISC International Digital Libraries Initiative, the University of Michigan and Cedars are collaborating to explore technology emulation as a method for long-term access and preservation of digital resources. The project will develop and test a suite of emulation tools, evaluate the costs and benefits of emulation, and develop models for collection management decisions about how much effort and resources to invest in exact replication within preservation activity.

Collection-Based Long-Term Preservation, 1998–

www.sdsc.edu/NARA/
This project, led by a group of San Diego Supercomputer Center scientists, aims to develop a persistent archive to support ingestion, archival storage, information discovery, and presentation of digital collections, including digital images. One of its premises is the importance of preserving the organization of digital collections (the information required to assemble the digital objects) simultaneously with the digital objects that comprise the collection.

Data Provenance, 1999–2003

db.cis.upenn.edu/Research/provenance.html
Funded by the Digital Library Initiative, this University of Pennsylvania project examines the com-

47. The British Library Research and Innovation Centre sponsored a study to examine "post hoc rescue of digital materials": Seamus Ross and Ann Gow, "Digital Archaeology: Rescuing Neglected and Damaged Data Resources," *eLib: The Electronic Libraries Programme, eLib Supporting Studies,* www.ukoln.ac.uk/services/elib/papers/supporting.

48. For example, German scientists recently examined documentation and interviewed retired technicians to recover digital data lost to software and hardware obsolescence in the East German archives: Gerd Meissner, "Unlocking the Secrets of the Digital Archive Left by East Germany," *The New York Times* (March 2, 1998): p. D5.

49. "Archiving and Distribution Working Group Charter," www.globalcollaboration.org/jtc1/igbt/adwg.html.

ponents of data provenance, which is crucial for the reliability of any kind of digital information. It attempts to clearly identify and describe issues related to developing tools for recording data provenance.

OAIS (ISO Reference Model for an Open Archival Information System)

ssdoo.gsfc.nasa.gov/nost/isoas/us/overview.html (under Reference Materials)
The OAIS reference model currently under review (as an ISO Draft International Standard) provides a framework for long-term digital information preservation and access, including terminology and concepts for describing and comparing archival architectures. It aims to facilitate a better understanding of the requirements for long-term preservation and access to provide a common framework. Cedars and NEDLIB are basing their preservation strategies on this model.

InterPARES Project (International Research on Permanent Authentic Records Electronic Systems), 1999–2001

www.interPARES.org
A multinational initiative, the InterPARES Project has the goal of developing the theoretical and methodological knowledge essential to the permanent preservation of electronic records. The research areas include identifying the electronic record elements that need to be maintained, developing criteria to appraise electronic records for preservation, and formulating principles for the development of international, national, and organizational preservation strategies.

Kulturarw³ Heritage Project, 1999–

kulturarw3.kb.se/html/projectdescription.html
The Royal Library of Sweden is testing methods of collecting, archiving, and providing access to Web-based Swedish electronic documents. The idea is to use robots to download all the Swedish Web pages automatically.[50] Although the project currently does not focus on preservation, it is growing into a broader Nordic initiative that may explore the long-term preservation of this archive.

---

50. Brewster Kahle and Peter Lyman are also testing this model in the US for their "Internet Archive": Lyman and Kahle, "Archiving Digital Cultural Artifacts."

Networked European Deposit Library (NEDLIB), 1998–2000

www.konbib.nl/nedlib
NEDLIB is a collaborative project of European national libraries to build a framework for a networked deposit library. Among the key issues it explores are archival maintenance procedures and the link between metadata requirements and preservation strategies. The project has adopted the OAIS reference model as a basis for a Deposit System for Electronic Publications (DSEP) and is developing a testbed for emulation-based strategies.

PRISM (Preservation, Reliability, Interoperability, Security, Metadata), 1999–2003

www.prism.cornell.edu
Cornell University's PRISM is a four-year project funded by the Digital Library Initiative to investigate and develop policies and mechanisms needed for information integrity in the context of a component-based digital library architecture. The key research areas include long-term survivability of digital information, reliability of information resources and services, interoperability, and security (the privacy rights of users of information and the intellectual property rights of content creators), and metadata that ensures information integrity in digital libraries.

Preservation Management of Digital Materials, 1999–2000

ahds.ac.uk
The UK's Arts and Humanities Data Service (AHDS) is developing a framework that will enable organizations to evaluate the long-term value of their digital collections and to assess the long-term access requirements. The project will also explore appropriate organizational and technical options for digital preservation, and identify their costs and benefits. The study will develop and test a decision-making tree to be used in cost-benefit analysis of digital preservation options.

PANDORA (Preserving and Accessing Networked Documentary Resources of Australia), 1996–

www.nla.gov.au/pandora
Among the achievement's of the National Library of Australia's PANDORA Project are establishing an archive of selected Australian online publications, developing several digital preservation poli-

cies and procedures, drafting a logical data model for preservation metadata, and outlining a proposal for a national approach to the long-term preservation of these publications. The project currently focuses on specifications for a technical infrastructure for storage, delivery, and management.

Preserving Cornell's Digital Image Collections, 1999–2000

www.library.cornell.edu/preservation
The goal of this IMLS-funded project is to plan and implement a digital preservation strategy for the Cornell University Library's digital image collection. Among the research objectives are to investigate the current and emerging file formats for long-term utility, to explore functional requirements for storage, and to develop a framework for requisite preservation metadata. One of the goals of this project is to assess the required staff, equipment, space, time, and finances for ten years of maintenance.

RLG-DLF Task Force on Policy and Practice for Long-term Retention of Digital Materials, 1999–2000

www.rlg.org/preserv/digrlgdlf99.html
The Research Libraries Group and the Digital Library Federation formed this task force to gather and analyze existing digital preservation policies for locally digitized materials, electronic records, and electronic publications. It will create one or more digital preservation policy frameworks that adequately address the material types and relevant institutional contexts.

Stanford Archival Digital Library Architecture: Archival Digital Libraries Repositories, 1999–

www-db.stanford.edu, link to Digital Libraries, then Projects, then Managing Information
The goal of this Stanford University project is to design and implement a scalable digital library repository (DLR) to ensure long-term archival storage of digital objects. The DLR will offer its clients a variety of services for managing intellectual property and for safeguarding information. Among the research goals of the project are identification of digital objects in a distributed and changing environment, the replication of digital objects for archiving, and the management of metadata.

9 Projects to Programs:

MAINSTREAMING DIGITAL IMAGING INITIATIVES

Anne R. Kenney

OVER THE PAST DECADE, libraries and archives have come to appreciate how well digital imaging efforts extend their reach. Money has been earmarked for such efforts, institutional rankings have been measured by things digital, and digitization has opened up major new means for making cultural resources available. User response to Web-accessible material has been overwhelming and positive. As we begin the 21st century, however, libraries and archives face a critical transition in which digital projects must give way to digital programs to survive. In other words, institutions must come to terms with the digital collections they develop. Projects by their very nature are of limited duration and scope, most often involving efforts to create digital resources. Programs are ongoing and encompass the full life cycle of digital resources, from selection and creation to management, access, and preservation. Programmatic efforts cannot be self-contained or viewed as separate from or parallel to the core institutional mission. To succeed, digital imaging programs must permeate the institutional culture and daily functions.

As institutions transition from projects to programs, digitization must be considered in a very broad context. We must strive to understand the technical environment in which these materials will be managed, accessed, and used, as well as the costs of maintaining digital collections beyond their current service objectives.

A STRATEGY FOR TRANSITION

The move from projects to programs is based on the premise that *digital collections are institu-*tional assets. Institutions must safeguard these investments to maintain their long-term value and utility. Cultural thinking must shift away from viewing digital imaging efforts as short-term or experimental.

You cannot move from projects to programs overnight so you must develop management strategies to address the transition. One issue to be faced is institutional resistance. Digital projects may enjoy a high profile at your institution, but regular staff can view them as flashy, ephemeral efforts that have little to do with your core mission. Although many are initiated by outside funding, digital projects ultimately require institutional resources to be sustained. As Don Willis and Drew Lathin argue in their sidebar to chapter 1, staff members may feel threatened by a competition for resources or by changes in the way things are done, or perhaps even become concerned that digitization threatens jobs. One senior administrator put it bluntly, "The staff will change...or the staff will change." Inevitably tensions arise as "cybrarians" and librarians learn to live in the same system. Although technical change is an inevitable consequence of this effort, overcoming institutional resistance may be the biggest challenge. John Secor argues that digital technology is "more an adaptive challenge than a technical challenge."[1] Strong administrative support, a sustained effort to communicate the relevance of digitization to the institution's mission, and staff involvement at all levels

1. John R. Secor, "Digitizing as Strategy," in *Economics of Digital Information. Collection, Storage, and Delivery*, ed. Sul H. Lee (Binghamton, NY: The Haworth Press, 1997), p. 104.

will be essential for viewing digital imaging programs as a legitimate part of doing business. The transitional period will not end until this occurs.

Even when institutions are committed to mainstreaming digital programs, management is difficult because there are no blueprints and few standards or best practices to follow. There is a growing body of literature on managing digital projects, but managing programs differs from managing projects.[2] It is also very different from managing established, ongoing programs that have clearly defined objectives and well-developed operational procedures. Digital imaging programs, like other initiatives involving the use of information technologies, often face major uncertainties about rapidly changing technology and underdocumented costs. Frequently, no precedent guides difficult decisions. In addition, digital programs will affect other operational units in ways that are difficult even to imagine. In effect, managing digital programs, especially in the transitional phase, combines project management, traditional management, and risk management.

Devote Institutional Resources to the Transition

A continuing series of digital projects does not constitute a program, and the shift in orientation requires careful planning and the commitment of both fiscal and human resources. To ensure the successful development of a fully integrated, supportable program, appoint a transition team with representatives from all potentially affected units. Get top administrative support for your team members, including reduced workloads for the duration. Depending on the size and extent of the digital imaging effort, you may need to hire a full-time transition manager and allocate funds to support the team's work.

Develop Formal Policies to Encompass the Life Cycle of Digital Resources

The development of formal policies is a prerequisite for establishing institutional programs and identifying resource requirements to support them. A number of institutions have developed policies for various aspects of digital imaging programs, such as conversion, selection, and copyright, but very few have created comprehensive policies that cover the full gamut of issues associated with managing digital assets over time.[3] The AHDS in the UK has undertaken some of the most promising work in this area. The need for a comprehensive approach to policy development—and an appreciation for the differences between analog and digital programs—is central to this

2. Recent publications devoted to managing digital efforts include Maxine Sitts, ed., *Handbook for Digital Conversion Projects: A Management Tool, Focusing on Preservation and Access for Paper-based Collections* (Andover, MA: Northeast Document Conservation Center, forthcoming); Lisa L. Macklin and Sarah L. Lockmiller, *Digital Imaging of Photographs, A Practical Approach to Workflow Design and Project Management*, LITA Guides no. 4 (Chicago: American Library Association, 1999); Susan Jephcott, "Why Digitise? Principles in Planning and Managing a Successful Digitisation Project," *The New Review of Academic Librarianship* 4 (1998): pp. 39–53; "Digitisation: A Project Planning Checklist," *Arts and Humanities Data Service (AHDS)*, ahds.ac.uk/resource/checklist.html; Peter Noerr, "The Digital Library Tool Kit," *Sun Microsystems*, www.sun.com/products-n-solutions/edu/libraries/digitaltoolkit.html; Sean Townsend, Cressida Chappell, and Oscar Struijvé, "Digitising History: A Guide to Creating Digital Resources from Historical Documents," *AHDS Guides to Good Practice*, hds.essex.ac.uk/g2gp/digitising_history/index.html. A number of sites offer a series of questions to answer or a checklist for planning for a digital imaging project, including Library of Congress, National Digital Library Program, "NDLP Project Planning Checklist," *Library of Congress, American Memory*, memory.loc.gov/ammem/prjplan.html; "Guidelines and Standards: Getting Started," *Colorado Digitization Project*, coloradodigital.coalliance.org/standard.html; "Questions to Consider Before Beginning an Image Database Project," *University of Illinois Library, Digital Imaging & Media Technology Initiative*, images.library.uiuc.edu/20questns.html; "Image Scanning," *Harvard University, Library Digital Initiative*, hul.harvard.edu/ldi/html/image_scanning.html; and "Questions for Consultancy," *TASI*, www.tasi.ac.uk/aboutus/proquest.html.

3. A notable exception is the California Digital Library, which has released standards for "archival quality" digital image collections that cover a full range of issues: "California Digital Library: Digital Image Collection Standards," (September 1, 1999) www.ucop.edu/irc/cdl/tasw/Current/Imaging.Stds-090199/Imaging.Stds-090199.pdf. Many institutions have adopted formal copyright and restrictions policies governing Web access, and AHDS maintains a "Rights Management Framework" (ahds.ac.uk/bkgd/rmf1.htm). The lack of formal digital archiving policy statements has motivated RLG and the Digital Library Federation to establish a task force whose work, including a policy framework for digital preservation, is to be completed in spring 2000 ("RLG-DLF Task Force on Policy & Practice for Long-term Retention of Digital Materials," *RLG*, www.rlg.org/preserv/digrlgdlf99.html). Columbia University has provided technical recommendations for digital imaging projects conducted by faculty, students, and staff at the university; Image Quality Working Group of ArchivesCom, a joint Libraries/AcIS committee, "Technical Recommendations for Digital Imaging Projects," www.columbia.edu/acis/dl/imagespec.html.

effort. Through several publications in the Managing Digital Collections series of AHDS, Neil Beagrie and Daniel Greenstein have developed a policy framework for digital collection management that consists of a number of closely interrelated and mutually dependent modules. These have been variously categorized but essentially cover data creation, selection and evaluation, management (including structure, documentation, storage, and validation), resource discovery, use, preservation, and rights management. The framework recently has been applied to AHDS's own digital collections and it informed the Joint Information Systems Committee/National Preservation Office Study on the preservation of electronic materials. The reports on these applications are most useful for institutions seeking to develop a

comprehensive set of policies and tying resources to them.[4]

Of particular importance in developing policies is the recognition that institutions have a stake in digital content creation. As discussed in previous chapters, how images are created can have a pronounced impact on how or whether—and at what cost—they can be effectively managed as long-

4. Neil Beagrie and Daniel Greenstein, "A Strategic Policy Framework for Creating and Preserving Digital Collections" (July 14, 1998), *AHDS*, ahds.ac.uk/manage/framework.htm; Neil Beagrie and Daniel Greenstein, "Managing Digital Collections: AHDS Policies, Standards and Practices, Version 1" (December 15, 1998), *AHDS*, ahds.ac.uk/public/srg.html; Tony Hendley, *Comparison of Methods of Digital Preservation. A JISC/NPO Study within the Electronic Libraries (eLib) Programme on the Preservation of Electronic Materials* (London: Library Information Technology Centre, 1998).

When the Ideal Meets the Pragmatic—or the Rubber Meets the Road
Sarah Thomas
University Librarian, Cornell University Library

In 1998, when the Cornell University Library developed a tactical plan to guide its activities, staff rejected the objective "Build a digital library." They preferred to incorporate digital initiatives into existing functional areas, observing that the expression "digital library" led to naïve interpretations that all content and services would be exclusively digital. The Cornell staff recommended subsuming "digital" under the general concept of "library." In actual practice, this is a complex assignment, since institutions rarely discard services voluntarily and budgetary increases to add new capabilities are few and far between. In the norm, external funding has powered the initial wave of innovative services in libraries. Often, the first generation of library systems has received substantial foundation support. Subsequently, agencies have contributed millions in grants for retrospective conversion. Large-scale microfilming and preservation projects continue to benefit from outside subsidies. Recently, digital efforts have been the rage.

If history is a guide, outside organizations will not sustain digital activity once it crosses over into the library mainstream. Ongoing operations are traditionally an institutional responsibility. Organizations integrating digital capabilities into their daily operations have several avenues for financing them. They can increase their base budgets by arguing persuasively that in the Information Age, the library warrants a larger share of university expenditures. They can campaign for endowment funds to underwrite the costs of informational technology, expertise, and new services. They can improve productivity by reengineering library functions. Lastly, and perhaps most difficult culturally, they can replace existing services with new models.

The hybrid model manifested in today's libraries is predominantly print- and hard-copy oriented and is managed in a very

labor-intensive manner. Despite an increasing investment in hardware, software, and technical staff over the past two decades, a relatively small percentage of Cornell Library's budget supports this area, perhaps less than 10%. Purchase of electronic journals, database access, full-text, audio, geo-spatial and visual resources represents approximately 10% of a research library's collection development budget. Obviously, if one believes that the 21st century will truly usher in the Digital Age, libraries should devote much more of their resources to digitization and its accompanying services. Demand for electronic content will grow. If readers are to mine the libraries' legacy holdings and search across the spectrum of intellectual achievements, libraries will need to construct solid bridges to lead them from the digital world into the physical library, or they will have to digitize the most heavily used portions of their holdings.

How can a library director lead this revolution? There are several potential alternatives. First, one can create a climate in which innovation flourishes by encouraging risk taking and experimentation and rewarding active involvement with digital culture. Exposing staff to digital issues and engaging them in interaction with expert speakers and real or virtual presentations on digital developments can increase acceptance of new ways of doing business. To foster greater facility with tools and digital content, the library should establish a staff-training program for structured learning. Supervisors can let staff "play" on the Web. More puritanical types may raise the educational content of this exploration by directing staff to particular sites and then engaging in discussion about their experience. The goal should be a pervasive proficiency in the construction and use of electronic resources, coupled with a sense of the need for life-

CONTINUED ON PAGE 156

CONTINUED FROM PAGE 155

long learning. Revised position descriptions and performance measures would underscore the centrality of digital literacy and expertise.

Libraries could hasten the transition to new electronic products and services by offering internally funded incentives. At Cornell, a small, competitive grant program financed by donor support resulted in the testing of several innovative ideas. As simultaneous advantages, staff gained experience, new products were delivered, and the profile of digital initiatives in the library grew. Other painless inducements to raise funds for digital activities are the solicitation of gifts or the earmarking of base fund increases off the top. For example, unrestricted gifts can be directed towards digital reformatting and archiving, and a percentage of materials budget increases can be reserved for the purchase of or licensing of digital content.

As digital tools and resources increasingly penetrate library operations over the coming years, they will function less as novelties. Library gateways will provide access to print and online materials through catalogs and indices, and metadata will link bibliographic surrogates with full text and other digital manifestations, such as images, sound, and data sets. Microfilm reading rooms will receive fewer onsite visits as microfilm is converted to digital form. Electronic reserves will be commonplace, but may be associated with the course Web site, with the library only handling copyright clearance or converting nondigital formats into electronic media. Online reference will overtake face-to-face queries and we may see automated distribution of reference queries by time zone to coordinated centers. Optimized technical services will further reduce staffing requirements. The decline in some activities and the rise in others will enable a reallocation of resources to digitization. Collaborative efforts with regional, national, and international partners will increase interoperability, reduce redundancy, and accelerate the transition to the new order. Only by working together will institutions be able to meet this extraordinary technical, political, and cultural challenge and seize the opportunity to move to a higher plane of understanding.

Many of these recommendations have an incremental quality, but the speed of users' adaptation to electronic resources and communication will accelerate implementation. The transformation will require dynamic planning, prioritization of services and funds, and a clear understanding of users' needs. To that end, libraries should extract a ranked order of current and potential services from surveys, focus groups, and other methods. This objective data can then be weighed against the cost of providing various services and the increased productivity they would yield for users or for staff. Despite a temptation to cling to the less popular services preferred by an adamantly vocal minority, intrepid leaders will keep ahead of the curve by augmenting a steady reallocation of funds with new monies derived from partnerships, external agencies, and gifts. Library services will change dramatically in the coming decades but the library's core responsibility will remain: linking users and information in support of learning and the creation of new knowledge.

term institutional assets. Enforceable minimal guidelines for content creation (for both images and metadata) should flow from the development of policies.

Tie Policies to Institutional Resources

Digital initiatives will not become fully implemented programs until resources are tied to them. To determine resource requirements, you have to be able to quantify and justify the cost of managing programs. This is difficult to do, since there is too little sustainable effort to point to for comparing budgets, staffing needs, and the like. Institutional buy-in, however, will depend on being able to predict the kinds of resources that will be needed over an extended period. Resources will divide into one-time expenditures and ongoing commitments. Senior administration must empower the transition team to work with division managers to assess impact, identify resources, and quantify efforts. The team should characterize digital collection management in terms of collection growth, the evolving technical infrastructure, and anticipated use.

Resource requirements should be justified by perceivable benefits or risks. Affected units should assess the impact of additional work for digital programs, rather than argue for parallel structures to accomplish it. For instance, technical services may have to establish a persistent naming service; reference will need to devote staff resources to serving networked users; archivists and collection development librarians will become concerned with technical capabilities for rights management. These additional demands should be quantified as much as possible in terms of staff time, training, user education, and technical requirements. The development of a strategic plan for image libraries at the National Museum of Australia provides an example of such an approach.[5]

5. Katy Bramich and Judith Cannon, *Capturing the Big Picture: Strategies for Image Libraries at the National Museum of Australia*, abridged version (National Museum of Australia, 1998), www.nma.gov.au/aboutus/products/bigpic.pdf. Simon Tanner and Brian Robinson, *JISC Image Digitisation Initiative: Feasibility Study (Final Report)* (Hatfield, Hertfordshire: University of Hertfordshire, Higher Education Digitisation Service, 1998), heds.herts.ac.uk/Guidance/JIDI_fs.pdf, details background, method, baselines, production processes, and potential costs for a range of document types.

Analyze Current Digital Imaging Projects for Efficiencies and Economies

While digital imaging projects require an institution to develop a workplan and identify costs, many projects have been hampered by unanticipated ramping up efforts or were not motivated by accountability.[6] Projects may lack productivity or quality controls or a complete reckoning of the full costs associated with the effort. In the transition phase, however, an institution should assess current practices with an eye towards increasing efficiency and effectiveness. Integrating business-oriented practices—such as workflow and systems analysis, knowledge management, business process analysis, cost–benefit analysis, and process architecture—into the development and management of digital assets should become the responsibility of the transition team.[7] Project management tools, such as Microsoft® Project and products from PRINCE Software Inc. may prove useful in analyzing digital imaging projects after the fact.[8] Other useful tools include decision matrixes covering various aspects of digitization.[9] The transition team should identify factors affecting production such as bottlenecks, task-sequencing dependencies, inconsistencies, redundancies, and unnecessary efforts. Digital imaging functions should be considered in the larger context of other digital resources and other IT-related operations, such as the library management system. Consider the benefits of centralizing functions, multitasking, adopting an incremental approach, and outsourcing particular tasks. As with preservation microfilming, many institutions have found it cheaper to outsource digitization than to maintain an in-house operation. Establishing a good working relationship with vendors is critical to this effort.

In addition to considering economies associated with digitization efforts, consider whether digitization can result in cost savings in other areas. For instance, can online access save money for circulation or shelf maintenance? Can you quantify benefits in terms of prolonging the life of vulnerable originals? Providing greater security? Can digitization replace some services, such as photocopying or interlibrary loan? Proponents of digitization have cited economic benefits, but institutions have had a hard time realizing cost savings. In a recent survey of six museums and research libraries, all responded that digitization and the creation of Web sites had not eliminated any traditional services.[10]

Document the Process and the Product

As part of conducting a systematic analysis of imaging projects, the transition team should assess the level and completeness of available documentation. Adequate documentation of digital resources is vital to their effective management, use, and long-term value. Information on processes and project decisions is also a prerequisite to developing a sustainable program—reconstructing this information after the fact can be resource intensive. An institution needs to invest in its staff to mainstream a digital imaging program, but it should not depend on particular individuals to provide the corporate memory when policies and processes have not been well documented.

6. Participants in the LC Ameritech projects described workflow problems in Library of Congress/Ameritech, "Lessons Learned: National Digital Library Competition," memory.loc.gov/ammem/award/lessons/workflow.html.

7. A good example of this assessment is the 1996/97 Depository of Netherlands Electronic Publications (DNEP) workflow project, which designed operation procedures for a sample of 100 electronic publications. All aspects of the workflow were considered, including selection, acquisition, registration, cataloging, access, storage, preservation, and search and retrieval; Trudi Noordermeer, "Deposit for Dutch Electronic Publications: Research and Practice in The Netherlands," in *Research and Advanced Technology for Digital Libraries: First European Conference, ECDL '97, Pisa, Italy, September 1–3, 1997*, Lecture Notes in Computer Science no. 1324, ed. Carol Peters and Constantino Thanos (Berlin; New York: Springer, 1997), pp. 361–373.

8. P. J. Day, *Microsoft Project 4.0 for Windows and the Macintosh: Setting Project Management Standards* (New York: Van Nostrand Reinhold, 1995); CCTA, *PRINCE2, Project Management For Business*, (London: The Stationary Office, 1997). C. Harvey and J. Press, *Databases in Historical Research: Theory, Methods And Applications* (London: Macmillan, 1996) as cited in Townsend, Chappell, and Struijvé, "Digitising History."

9. See in particular Oxford University, "Decision Matrix/Workflows," www.bodley.ox.ac.uk/scoping/matrix.htm, and "Selection for Digitizing: A Decision-Making Matrix," *Harvard University Libraries, Preservation*, preserve.harvard.edu/resources/digitization/selection.html.

10. One respondent indicated that CD-ROM versions of some journals might be eliminated with online access, but this has not yet been done. John Chadwick et al., "Assessing Institutional Web Sites" (report prepared for the Council on Library and Information Resources, September 30, 1999), p. 11.

Outsourcing Digitization: A Service Provider's Perspective
Meg Bellinger
President, Preservation Resources

As more institutions embark on ambitious digital imaging initiatives, they are contracting all or part of the work to service providers. Many institutions are relatively new to digital imaging and many service providers are unfamiliar with the particular needs of cultural institutions. Because they lack a common base of understanding, the relationship is often difficult. A successful client–vendor relationship depends on a mutual and standardized vocabulary, clearly defined and measurable project goals, and an understanding of the capability and limitations of the technology.

The familiar models that have served so well for large-scale microfilming projects are not adequate for digital reformatting projects. In preservation microfilming, the creation of the medium is an end in itself, with a predictable use of the content and few variables. Digitization is about access, not media. Image capture is only an initial step in a complex process that should be informed by the end uses. Display and printing requirements affect the resolution of image files and of their derivatives. Network access affects the file format and size. Operating systems and image databases affect file naming, organization, and delivery.

For digitization projects, institutions and service providers are working with developing technologies and a new vocabulary, creating new quality and production benchmarks, and trying to determine best practices. All the while, digital technology continues to evolve. Both parties must collaborate to determine capture requirements, costs, and deliverables; manage the process; and agree on criteria.

From a service provider's perspective, the following recommendations will assist institutions in reducing potential problems or misunderstandings.

Qualify the Service Provider
Be certain that the digital imaging service provider you select is experienced and knowledgeable about the concerns of cultural institutions. Reputable businesses do not intentionally overcommit, but they may not understand the challenges posed by scanning research materials. Is the service provider prepared, for example, to treat the original artifact carefully and handle a variety of documents and bibliographic anomalies, non-English languages, and inconsistent numbering systems? Most production scanning equipment, management software, and procedures were developed for uniform, routine material such as business forms. A service provider may have to adapt out-of-the-box technology or create customized solutions for cultural institutions.

Determining project goals and parameters involves a client–vendor dialog. Some service providers have developed a questionnaire or checklist for institutions to clarify project requirements as well as determine products and costs.[1] This dialog is necessary to strike the right balance between cost-effectiveness and the quality or customization of the product.

Send document samples and request digital samples of work to ensure clarification of expectations and capabilities. Depending upon the size of the project, expect some cost for the time it takes to prepare the sample and a written analysis. Do not forego this step. A small amount of money spent up front can save considerable amounts later.

Develop a Detailed RFP
Your RFP should clearly define content and expectations but also allow the service provider enough flexibility to develop recommendations based on actual outcomes. As a starting point, see the *RLG Guidelines for Creating a Request for Proposal for Digital Imaging Services*.[2] Difficulties arise when too many assumptions are made or the sample is too small. One set of scanning specifications or generalized document descriptions rarely covers everything in a large collection of diverse materials. Too often we find that clients are not familiar enough with the contents of the project, are unclear about the end uses, and have not decided on desired bit-depth and resulting file sizes. These will affect special handling, schedules, and the cost of media or other mechanisms for delivery.

We have yet to achieve the kind of project consistency that will allow us to have one set of prices. Until standard practices and delivery expectations apply to consistent types of documents (for example, European-language monographs versus Slavic pamphlets), we can only base cost estimates on deliverables—the time calculated to perform bibliographic inspection, scanner set-up, scanning, postscan processing and editing, quality assurance, metadata creation, file naming, additional quality assurance, and delivery to the system. Of the nearly 100 imaging proposals we have written to date, no two have been the same and we do not expect this custom approach to change soon.

The following increase project costs:

- Requirements for manual postscan processes, including editing, naming, and enhancements requiring manual work

- Nonuniform material

- Increasing bit-depth and, therefore, file size

- Delivery media and overall file size

- Selective rather than comprehensive scanning (partial reels or parts of texts)

- Multiple requirements, such as different bit-depths for illustrations versus text

- Specifications that require close analysis of material content

- Absence of technical information for microforms

Pricing for each separate function is useful for negotiation and for planning what functions can occur elsewhere or later.

CONTINUED ON PAGE 159

CONTINUED FROM PAGE 158

Launch the Project with a Pilot Phase

A pilot phase tests the communication and understanding of expectations under the project conditions. Expect to repeat the process every time expectations or content changes. For example, even with a very successful relationship with the Library of Congress, for every National Digital Library collection, we evaluate the content and provide samples and recommendations for bit-depth and resolution. We then test the recommendations in a pilot phase. Some qualitative issues may be difficult to anticipate within broad initial requirements and the product needs to be checked for quality and characteristics such as the presence of borders from film scanning, how halftones look as bitonal images, and lightness or heaviness of text. Scanning deconstructs a collection and reveals characteristics that cannot be anticipated unless every page is preevaluated.

The pilot provides ramp-up time for both the institution and the service provider to address final details before the project reaches full production:

▶ To ensure that viewing software and devices are comparable so that results can be verified at the institution.

▶ To evaluate scanning and output requirements and naming or metadata requirements if necessary.

▶ To define rework and file reintegration issues, including tracking rework on multiple images.

▶ To set the framework for review. Bit-mapped images take up valuable space; be prepared to review and accept or reject quickly.

▶ To define procedures for deleting all iterations of the file.

Work Together

Cultural institutions should value the contribution of time and creativity a business makes to meet their needs. These efforts distinguish a service provider from its competitors. To convey one business's creative solutions to another means a real loss and great frustration for the business that invested time and energy. Creativity and development are necessary and desirable costs of business; cultural institutions should give due consideration to these costs.

Given the unique and valuable materials that cultural institutions hold combined with the developing nature of both technology and best practices for digital imaging projects, both parties need knowledge, understanding, and communication. The most successful relationships accept and build upon the intrinsic differences between businesses and cultural institutions and clearly define roles within each organization to mediate these differences.

Notes

1. See, for instance, "Preservation Resources Scanning Questionnaire," *Preservation Resources*, www.oclc.org/oclc/presres/scanning/scanquestion.htm.

2. "RLG Tools for Digital Imaging," *RLG*, www.rlg.org/preserv/RLGtools.html.

Undertake the Doable

In a presentation to the 1992 annual meeting of the Research Libraries Group, Don Waters suggested a number of principles for viewing the challenges of imaging technology. Among them were the KISS (Keep it Simple, Stupid) principle and the adoption of an incremental approach.[11] These two points are still valid today. As you move from projects to programs, avoid the temptation to promise more than can be delivered given the collective institutional energies and knowledge base. For instance, as we have noted, if your institution is scanning its map collection, you should be realistic about what can be delivered to users at their desktops. Or if you choose to provide access to the world, user demand may outstrip your current infrastructure. It is much preferable to scale up based on solid performance than to scale back because you cannot cope with an overly ambitious program proposal. Adopting an incremental, scalable approach is a sound business practice that applies especially in programmatic development involving the unknown.

Staff the Program with Well-Trained People

In his final report on Yale's Project Open Book, Paul Conway documented the impact of training and practice on processing costs. His extensive analysis showed that the "practice effect" improved productivity 44% for scanning and 50% for indexing.[12] Too many imaging projects rely on inexperienced or short-term staff, such as students, to perform critical tasks that affect the quality and costs of other functions. For instance, if inexperienced staff produce poor-quality images, the time spent on inspection can skyrocket, affecting workflow in other areas. Outsourcing offers trained staff, among other advantages. For those functions performed in-house, invest in staff-training programs to promote digital literacy and expertise and provide incentives to retain experienced staff members, particularly in critical IT-related areas. As John Price-Wilkin has noted, technical skills "are an important commodity, and building a production organization by

11. Don Waters, "Electronic Technologies and Preservation," *CLIR, Commission on Preservation & Access*, www.clir.org/pubs/reports/waters/waters2.html.

12. Paul Conway, *Conversion of Microfilm to Digital Imagery: A Demonstration Project* (New Haven, CT: Yale University Library, 1996), p. 18.

'leasing' such expertise is not realistic when the systems of access or management depend so fundamentally on them."[13]

Incorporate Flexibility

The transition team must serve as a change agent in two ways. It must ensure that digital imaging projects turn into programs that in turn remain flexible and open to alteration. It is responsible for

charting a course with the full understanding that some parts will remain viable for only a short time. We want to make informed decisions, understand the consequences of choices, respect the limitations of current practice and technology, and be prepared to right the course occasionally as things change. Some modifications will be brought about in response to changing user behavior and needs, and institutions should build an ongoing evaluation component into program development.[14]

13. John Price-Wilkin, "Moving the Digital Library from 'Project' to 'Production'" (paper presented at DLW99 in Tsukuba, Japan, February 1999), jpw.umdl.umich.edu/pubs/japan-1999.html.

14. Robert Rieger and Geri Gay, "Tools and Techniques in Evaluating Digital Imaging Projects," *RLG DigiNews* 3, no. 3 (June 15, 1999), www.rlg.org/preserv/diginews/diginews3-3.html.

Production Tracking

Paul Conway
Head, Preservation Department, Yale University Library

Creating a digital product can be quite an elaborate process. Today, most imaging initiatives are conducted as sequenced, self-contained projects, even when an institution has implemented an ongoing operational program. Project tracking establishes a system to gather and analyze information about the imaging production process, the source materials and digital files, and the characteristics of the final products. Production tracking and workflow analysis are fundamental to the long-term health of an imaging program in any library or archives.

Why Track Production?

If you contract any part of the production process to a service bureau, production tracking is perhaps the most direct way to know that the terms of the contract are being met. If a single administrative unit covers all aspects of your imaging process, workflow analysis is the principal mechanism for improving efficiency and effectiveness incrementally. Documenting your processes and procedures helps ensure the reliability of your products.

The information gathered from a production process shows that you can complete a project on time and within budget. Nonetheless, it is a rare funding proposal today that can point to the purposeful development of a production process and a budget based on the systematic analysis of production costs. Tracking information from small pilot projects can provide crucial information for evaluating costs and workflow procedures to develop larger-scale projects or ongoing programs.

Imaging initiatives in libraries and archives are particularly challenging and complex, due in part to the complexity of the source materials and in part to the fact that relatively few systematic studies of the digital imaging process have been published.[1] The information from a production tracking system is the raw material needed to develop a shared understanding of best practices by the library and archives community.

Who Should Develop the Tracking System?

The production manager should establish a tracking system in close consultation with the staff engaged in the production process. The tracking system should be designed from the perspective of those who will use the information for reporting, fundraising, and administration. Outside contractors should also be involved if information will be exchanged between contractors and the production manager. A number of service bureaus are encouraging institutions to develop joint production-tracking systems.

What Should a Tracking System Cover?

A production-tracking system that satisfies a variety of administrative needs and supports analysis, interpretation, and generalization requires rich data about the production process, the objects involved, and the overall administration of the effort:

Production Process

- Queue (number of items, unique identifiers, physical location)
- Pipeline (number of items at various processing stages)
- Status tracking (to pinpoint the location of any item in the pipeline)
- Project tracking (to assess how much has been completed)
- Beginning and ending dates and times for each object converted
- Time spent on each major step of the process for each object or for a sample of the overall production
- Bottlenecks and information on how they are resolved
- Personnel (time spent on various tasks, log-in and log-out)

CONTINUED ON PAGE 161

Other changes will result from technological developments, and digital programs must incorporate the means to monitor and forecast technological change so as to assess risks to collections but also prepare for new ways to access and manage digital resources. The proposed digital preservation strategy outlined in chapter 8 requires that an institution assign staff members to serve as technological scouts. Changes both big and small are inevitable; in his sidebar, Cliff Lynch predicts four major technological developments that will affect digital imaging programs in the years to come.

Develop Institutional Programs in the Context of Multi-Institutional Collaboration

A recent international survey reported that 85% of respondents had cooperated with other organizations in developing digital efforts.[15] Certainly there are good examples of collaborative projects undertaken around the world—the UK-based Internet Library of Early Journals, the Australian Ferguson

15. Sara Gould and Richard Ebdon, ed., "IFLA/UNESCO Survey on Digitisation and Preservation," *International Preservation Issues* 2 (1999): pp. 14–16.

CONTINUED FROM PAGE 160

Objects

- Characteristics of the original objects (physical size, format or genre, color ranges)
- Characteristics of the digital objects (pixel dimensions, pixel depth, file format, file size)
- Scanner settings and routine enhancement procedures
- Versions and derivatives (pixel dimensions, pixel depth, file format, file size)
- File names (full directory paths) for all versions
- Unforeseen technical challenges

Administration

- Invoice tracking
- Accounting for administrative overhead
- Amortized capital costs (for equipment and space)
- Equipment and network costs
- Maintenance and leases on hardware and software
- Production capacity of scanning equipment (minimum, maximum, typical daily production)
- Total time major pieces of equipment are in use
- Maintenance and equipment downtime

Usually, production-tracking information is gathered on one or more forms and then analyzed quantitatively and qualitatively. Information gathered to support a tracking system also supports other imaging process needs, particularly the generation of administrative and structural metadata.

Increasingly, in large-scale imaging production operations the software that manages the image database and production system automatically generates most (but not all) important production-tracking information. In business and industry, workflow software routes digital objects through a production system, identifies production bottlenecks, and keeps track of locations, names, versions, and other nettlesome details.[2]

When Should Production Tracking Be Instituted?

Establish your tracking system during the design of the production process. System design must begin with an agreement on how the information from the tracking system will be reported, analyzed, and used by the production manager as well as by senior administrators. Imagine a series of charts and graphs or a narrative report containing specific conclusions about the accomplishments of the project. Then identify the data you'll need to produce such a chart or report, as well as the descriptive and interpretive statistics that will strengthen your report. This approach is better than designing workflow tracking forms and then trying to figure out later how to make sense of the information you have gathered.

Who Should Have Access to the System?

One purpose of production tracking is to identify and solve bottlenecks in the workflow before they threaten the success of the program. The production-tracking system, therefore, should be accessible in all locations where aspects of the digital imaging process occur, including the digital imaging production area and units that handle indexing and metadata creation.

Contractors should feed information into your production-tracking system. Although the contractor's detailed production information may not be necessary, the mutual exchange of status-tracking information and cost/timing data can increase the overall confidence of the production managers at both your institution and the contractor's. It might be useful to obtain from the contractor a breakdown of costs of specific subprocesses to help you assemble information on the overall cost of the production process and determine if outsourcing would be effective and efficient for similar efforts.

Notes

1. Paul Conway, "Yale University Library's Project Open Book: Preliminary Research Findings," *D-Lib Magazine* (February 1996), www.dlib.org/magazine.html; Conway, *Conversion of Microfilm to Digital Imagery: A Demonstration Project* (New Haven, CT: Yale University Library, 1996).

2. Bill Chambers, Richard Medina, and Kelley West, "Drive the Revenue with Workflow," *Imaging & Document Solutions* 8 (February 1999): pp. 18–30; Penny Lunt, "New Ways to Flow Your Work," *Imaging Magazine* 7 (January 1998): pp. 64–73.

Cumulative Evolution as Revolution:
Four Trends that Will Change the Rules for Digital Imaging Programs

Clifford Lynch
Executive Director, Coalition for Networked Information

Digital imaging is technology intensive: there never seems to be enough bandwidth, storage, capture speed, or pixel resolution. What is economically and technically possible evolves substantially from one year to the next. As we try to understand how future developments will change the rules for digital imaging programs, we focus perhaps too narrowly on technology. We tend to look for revolutionary technological change and constantly struggle to filter the hype promising such developments. I propose that four developments will significantly change the way we approach digitization programs in the next few years. Rather than breakthroughs, these changes reflect evolution creeping up on us.

Technology Price/Performance Improvements

We know technology price/performance improves steadily, but it is hard to measure the rate of change and even harder to recognize when cumulative incremental improvements mandate fundamental rethinking of our approaches. For example, near-line storage systems (tape robots, juke boxes, etc.) have caused many data-management headaches for imaging projects. These specialized technologies represent a niche market and tend to be unreliable, temperamental, and *slow*; and they are usually not well integrated with other software, such as database-management systems. As disk storage prices continue to drop, more and more projects will be able to operate entirely online, avoiding all the complexities and operational problems of near-line storage systems. Certain applications—for example, remote sensing—may always exceed online storage capabilities, but these will increasingly be specialized exceptions rather than the rule. For still images scanned from manuscripts and photographs, it will not be long before we simply order enough disk space to accommodate the databases.

While lossless compression techniques will have their place for some time to come due to the redundancy in most images (and thus a high compression payoff), at least for storage purposes we will see less interest in lossy compression schemes, such as JPEG, based on fooling the eye. Increasingly, the major applications for lossy compression will be limited to streaming media such as audio and video.

The implications for network bandwidth are more confusing. Bandwidth for users in university offices, computer labs, libraries, and dorms will grow enormously and elaborate multiresolution transmission schemes for images will become less and less important. Thumbnails will still be popular because they make for good browsing, not because of limited bandwidth. At the same time, lots of users—at home or using wireless connections—will still have constrained bandwidth. Supporting these users well will exact a high price in extra system complexity.

It is easy to get locked into assumptions about the need for complex solutions driven by technology constraints, and it will be important to reexamine these assumptions aggressively every year or two, looking for ways to simplify our systems.

Imaging Technologies Becoming Imaging Systems

Today, imaging products do not fit smoothly together into a system. For example, a great deal of metadata can be automatically generated at the time of image capture and automatically transferred from the capture device along with the images. Imagine if the entire processing chain—scanners and cameras, monitors, and processing software—automatically captured, maintained, and used color-calibration information. Imagine if metadata about capture settings, time, place, and other information were acquired as part of image capture and propagated automatically. This may come as digital cameras and scanners become commonplace consumer and low-end business items and as consumers produce more digital images than prints. These changes will reduce the costs of digital imaging programs, particularly when program planners adopt a consumer/commercial standard rather than insisting on better but enormously more expensive custom administrative metadata.

Or, consider something more speculative: The general public is now coming to understand just how easy it is to manipulate digital images. Given the use of imaging in areas such as law enforcement and news reporting, and the role of images as evidence in any number of legal proceedings, it is not unthinkable that we will see the development of "witness cameras." These systems could employ timestamps, global positioning system (GPS) receivers, wireless communications, digital watermarking, and an infrastructure of digital notary services to produce a signed, nearly tamperproof stream of images that include documentation about when and where each image was captured.

We need to be prepared to exploit standards and technology that make imaging products into systems. We also need to recognize that the "born-digital" images that enter our archival collections will start to incorporate such metadata.

Image Collections Are Networked
Information Resources

We are still thinking of corpora of digital images as if they were print-based collections; that is, people retrieve and examine materials, and then perhaps write about them, quote from them, or cite them. While this undoubtedly will not change, we will also see a major new development: people will link to digitized content in order to integrate it into new networked information objects. Scholars constructing Web sites that are descendents of monographs and encyclopedias will want to connect them to primary information; curators and archivists will want to assemble logical archives out of the materials housed in digitized collections at a number of institutions.

Digitizing projects will have to devote intense attention to developing long-term naming schemes for the digitized materials that can be used to make persistent links among digital objects that are created and managed by a wide variety of par-

CONTINUED ON PAGE 163

Project, the Nordic digital newspapers project, and the California Digital Library to cite a few. Many of these collaborative projects have been encouraged or required as a prerequisite of external funding. At the programmatic level, however, the value of collaboration from an institutional perspective gets too little attention and frequently there is insufficient recognition that important and complementary work is being conducted elsewhere. Collaboration is essential to ensure consistency across digital resources and avoid redundancies in such areas as selection, technical development, access, and user services. In the analog world, where a library's prestige has been measured by the size of its collection, such redundancies were considered part of doing business. In the digital realm, where enhanced access is emphasized over ownership, such redundancies represent little beyond added institutional expenses. Further, economies of scale are not sufficiently explored in areas such as data management or collaborative licensing, despite common rhetoric and growing evidence that individual efforts are not cost-effective. Enduring models will require true collaboration, transcending competitive instincts. Deanna Marcum's sidebar highlights the critical role that collaborative arrangements must play in developing sustainable institutional programs.

Two Case Studies

Many institutions are beginning the shift from a project orientation to digital programs and their experience demonstrates the value of careful planning and integration. Oxford University is at the beginning of this transition phase and the University of Michigan is well along the path toward full program development.

The Andrew W. Mellon Foundation funded a feasibility study on the development of a digital library service at Oxford University. The study's intent was to position Oxford in the transition from projects to services by consolidating existing initiatives and establishing the Oxford Digital Library Services (ODLS) to coordinate the creation, management, marketing, and delivery of digital surrogates. The August 1999 report recommends the appointment of a development team to oversee a three-year phased implementation. A director of ODLS would be charged with responsibility for four units—administration, data management, systems, and digitization service—and a staff of 21.5 FTE, to be built through recruitment and/or redeployment of existing staff. The full costs for the first three years, including staff, set-up, maintenance, and overhead, are projected to be £2,860,831, or $4,727,809.[16] A detailed five-year business plan covers both on-demand digitization and systematic digital collection building. The business plan posits 12 staff delivering 400,000 images per year. Funds for this service are expected from three sources: the university, products and services, and external support (estimated at £2 mil-

16. Stuart D. Lee, "Scoping the Future of the University of Oxford's Digital Library Collections" (September 1999), www.bodley.ox.ac.uk/scoping/report.html. (Exchange rate calculated at £1 = $1.6526 US.)

CONTINUED FROM PAGE 162

ties. Persistent names (URNs) deployed networkwide will need to be an integral part of any truly useful digitization program.

Descriptive Metadata and Image Retrieval

As we gain more experience, we are coming to recognize that creation of descriptive metadata is usually both critical and costly. Digital images are not very useful without retrieval tools. We need to view image description and retrieval more as engineering than as abstract philosophy. We do not understand this technology nearly as well as we understand capturing and calibrating pixels; in particular, we lack good real-world measures for cost-effective descriptive practices. We still struggle with unreasonable expectations, perhaps, for image description; we no longer expect bibliographic cataloging to allow us to find all books that interpret or elaborate the story of Faust, yet we sometimes at least dream of the ability to identify images that depict allegories.

With initiatives like the Dublin Core, we have conclusively abandoned the orthodoxy of bibliographic cataloging (or any other approach) as the one true model for image description. We can do image retrieval by content, though the languages are alien (to find pictures of meadows of flowers, one might ask for pictures with blue on the top and green on the bottom, with lots of bits of yellow in the green). Perhaps the goal of image-retrieval systems may be to provide a set of images for review, not all of which will be relevant to a query. For images of textual materials, optical character recognition (OCR) provides yet another approach for retrieving materials. Indexing algorithms—be they based on content recognition or OCR—are going to steadily improve. Deciding when to periodically reindex collections to take advantage of these improvements will be part of digital collection management. Image retrieval, and the creation of metadata to support it, will likely become a very cost-conscious engineering discipline, which may in the future determine what projects are and are not economically feasible.

Digital Barn Raising

Deanna B. Marcum
President, Council on
Library and Information Resources

In agrarian America, barns—at the center of the family farm—were built by the entire community. The barn was so crucial to the viability of the farm that it could not be built slowly or piecemeal. Barn raising comes to mind when thinking about the development of digital libraries, because such resources require a community effort, albeit a sustained one. A number of independent projects are under construction and we are learning a great deal about technical requirements and costs from them, but until these projects contribute to a broad vision, they will not result in a national digital library.

From the end of the 19th century until roughly 1990, libraries have worked to centralize and unify practices. Common classification and cataloging systems allow relatively easy navigation through collections of many different libraries, and a nationwide bibliographic system has made it possible for scholars located anywhere to consult the holdings of libraries everywhere.

Web-based digital resources, though a great step forward in offering an ever greater array of information, may present a great step backward for users' understanding and utilization of the system. We have an opportunity to create a national virtual collection that will offer enormous benefits to users, but we cannot achieve a fully effective national digital library without meaningful collaboration on standards, systems architecture, and search and retrieval systems.

Digital library projects that have been developed by universities are not now interoperable. Even though traditional libraries are standardized enough that a researcher can walk into any American university library and feel reasonably able to do research, there is no digital equivalent. Each project has been carried out in a different way. We have no catalog of digital resources. "Web-based" simply means users are free to cast about amid the chaos and find what they can. There is no unified place, easily recognizable and entirely functional, where researchers can go with reasonable expectations of finding a significant body of appropriate materials. This situation is replicated in other countries and cross-border collaboration has received very little attention.

If the ultimate goal is to build a national or international digital library, individual efforts amount to small, stand-alone sheds, to return to the metaphor of the barn. A large collection, easily navigated by anyone, anywhere, is possible only if we agree on the larger goal that is in the public interest. It means individual institutions that have a lot of justifiable pride in the projects they have created will have to forego some of their creativity and conform to a community-defined system.

CONTINUED ON PAGE 165

lion over the first five years). The university is committed to taking the recommendations forward as quickly as possible.[17]

One of the furthest along in the shift from projects to production services is the University of Michigan. Formed in 1996 through a commitment of resources from four campus organizations, the Digital Library Production Service (DLPS) headed by John Price-Wilkin has achieved a maturity consistent with a full-fledged program. In 1998, funding for DLPS was extended and consolidated as base budget resources. The program's success can be measured a number of ways:

▸ DLPS has been integrated into the library system and its program is viewed as consistent with the institution's mission to facilitate access to important information sources for research and teaching.

▸ Online availability has generated renewed interest in historical resources that were largely unused in their physical form.

▸ The production organization emphasizes the creation and management of long-term digital assets over short-term efforts.

▸ The library has invested in a highly skilled permanent staff, currently 20 FTE.

▸ DLPS has secured the support of a wide variety of on-campus projects to build a unified database primarily by demonstrating a cost-effective, highly functional host service rather than relying on administrative mandates.[18]

FINANCING DIGITAL IMAGING PROGRAMS

A major part of the transition from projects to ongoing programs will focus on determining the bottom line: What is it going to cost the institution? While there are still many unknowns, institutions should not expect to fully recover the costs of developing digital imaging programs.

17. R. P. Carr, director of University Library Services and Bodley's Librarian, letter to Anne R. Kenney, October 7, 1999.

18. Price-Wilkin, "Moving the Digital Library."

Costs of Digitization

There is no consensus on what it costs to create digital image files, much less maintain them and make them accessible. The available figures vary tremendously with the types of material being scanned, the image-conversion requirements, the hardware and software used, and the range of functions covered in the calculations. Unlike what has grown out of other conversion efforts such as preservation microfilming, no consistent price schedule applies to outsourcing image conversion from vendor to vendor or even from project to project. Meg Bellinger's sidebar discusses some of the reasons for this.

We probably know the most about bitonal scanning of disbound volumes at 400–600 dpi, with scanning estimates ranging from $0.10 to $0.30 per image for large production projects.[19] Figures for bound-volume scanning are higher, often up to twice that amount. Some institutions have found that they can obtain a better product and faster production rate when bound items are rendered into single leaves for scanning, even with the costs of rebinding.[20] However, improved capabilities for bound-volume scanning may soon reduce this difference.[21]

Although production rates for film scanning are theoretically very high, in practice current limitations have imposed difficulties that have reduced

---

19. These figures have been reported by Cornell, Michigan, and JSTOR. Also see Andrew Odlyzko, "Economics of Electronic Publishing: Journals Pricing and User Acceptance, The Economics of Electronic Journals" (paper presented at ARL's Scholarly Communication and Technology conference, Emory University, April 24–24, 1997), *Association of Research Libraries*, arl.cni.org/scomm/scat/odlyzko.html. Even lower costs are possible if pages can be sheet-fed. The Andrew W. Mellon Foundation has funded a project at the University of Michigan to document the full range digitization costs in a production environment, with results to be available in late 2000.

20. Ross MacIntyre and Simon Tanner, "*Nature*—A Prototype Digital Archive" (paper presented at the sixth DELOS Workshop Preservation of Digital Information, June 17–19, 1998), heds.herts.ac.uk/HEDSinfo/Papers/HEDSnature.pdf; "Internet Library of Early Journals (January 1996–August 1998) A Project in the eLib Programme: Final Report" (March 1999), www.bodley.ox.ac.uk/ilej/papers/fr1999/fr1999.htm.

21. NYPL reports production rates of 350 pages/hour using the double-page function of the Zeutschel M-5000 in grayscale mode. Beta-testing the new Minolta™ PS™ 3000 with 400-dpi, 8-bit capture capabilities, Cornell staff could scan and postprocess 125 pages/hour; the Center for Retrospective Digitization, Göttingen State and University Library, scanning at 400-dpi, 1-bit with both Zeutschel and Minolta scanners, averaged 250 pages/hour.

CONTINUED FROM PAGE 164

If all of the creators realize that they are contributing to a national digital library, they will pay more attention to interoperability and convenience for the user.

Several groups within the US are involved in activities that will provide the infrastructure for digital libraries:

▸ The 23 institutions that participate in the Digital Library Federation (DLF)—among them the most active in digital library activities—are struggling with the issues that other institutions will inevitably face as they become more involved in their own digital efforts. The subcommittees of the DLF are concentrating on standardized systems architectures, metadata, and digital archiving. By concentrating on standards, common protocols, and best practices, the DLF partners are defining how digital libraries should look and behave.[1]

▸ The Committee on Institutional Cooperation (CIC), in its work on a virtual electronic library for the member institutions, has developed a single interface for searching member institutions' online catalogs and electronic resources.

▸ The Coalition for Networked Information (CNI) has been active in exploring systems of authentication and authorization that will create circumstances in which users can locate and retrieve digital documents from a future national digital collection.

There are several parallel international efforts:

▸ The National Science Foundation and the European Union have initiated a series of working groups to address digital library development.

▸ The Networked European Deposit Library (NEDLIB), a collaborative project of European national libraries, has as its goal the development of the basic infrastructure upon which a networked European deposit library can be built.

▸ The European Visual Arts Project (EVA) focuses on providing networked, integrated access to European archives.

Some of the best systems minds in the US and around the world are addressing important technical issues, but a great need still exists. It is important that the library philosophers meet and declare the highest-level specifications for the digital library of the future. They must concentrate on how the digital library needs to be configured so that it continues to serve the broad and diverse needs of users. It is the vision that has yet to be articulated. Many members of the community stand ready to assist with the "barn raising," but a thoughtful blueprint is urgently needed.

Note

1. Digital Library Federation home page, www.clir.org/diglib.

scanning rates considerably. Today, the costs of film scanning remain equal to or higher than the costs of paper scanning at the same resolution and bit-depth. In a recent project to convert preservation quality film, Cornell paid nearly twice what it would have paid a vendor to do single-sheet scanning. The Internet Library of Early Journals (ILEJ) project found that costs for microfilm scanning were higher and the quality lower than direct bound-volume scanning.[22]

The speed of grayscale capture will soon come to rival that of bitonal scanning for some categories of material. For color scanning, however, the time and costs increase significantly—up to two to three times.[23] Figures for scanning graphic materials are an order of magnitude higher than for scanning text. Steve Puglia of the National Archives conducted a comparative analysis of digital imaging costs and his findings represent a sobering reminder that imaging is not inexpensive. For example, the National Archive's Electronic Access Project (EAP), which included manuscripts as well as graphic and photographic materials, achieved a daily production rate of 200 images for 10-MB files, with the scanning averaging $7.60/image.[24] JJT Inc. scanned photographic materials for LC, creating 18–20 MB image files, and was able to exceed the contract specifications of 525 images/day once in production mode. They used two digital cameras onsite and postprocessed and inspected the images in Austin, Texas.[25] Scanning figures go up for high-end imaging projects of museum holdings: the reported production rates to create 70–100 MB files range from a low of 15 images/day to a high of 70 images/day.[26]

If consistent figures for the actual digitization are difficult to obtain, there is even more divergence in calculating the total cost an institution will expend per image. Comprehensive figures are published less frequently, partly because it is less clear what to include. Obvious additional expenses are selection, materials preparation, quality control, file conversions, equipment, and infrastructure to mount files. But differences in consideration of items such as institutional overhead, staff contributions, cost sharing, and training and support could result in major discrepancies in reported expenses. In addition, metadata costs are extremely variable, depending in part on the nature of the materials being converted and the project's ambition to enhance access beyond analog materials. Although figures vary from project to project, Steve Puglia suggests that digitization typically represents one third or less of the total per-image cost, with the other two thirds going to metadata creation, administration, and the like.[27]

A Model for Calculating Costs

Full digitization costs are frequently underestimated or underreported. Consider for instance the various components that could be included in estimating budgets for image acquisition, as outlined in the *RLG Worksheet for Estimating Digital Reformatting Costs*.[28] The following assessment

22. "ILEJ Final Report." Including the microfilm, the costs were double those of bound-volume scanning. Without the microfilm, the price per image for film scanning was nearly $0.50 US (exchange rate £1 = $1.6526 US); the Early Canadiana Online project reported digitization costs of $0.44 US per image from microfiche ($1 US = $1.5257 Canadian).

23. Grayscale and color production figures reported in projects at the Library of Congress, the Smithsonian, the Beinecke Library at Yale University, and New York Public Library.

24. Steven Puglia, "The Costs of Digital Imaging Projects," *RLG DigiNews* 3, no. 5 (October 15, 1999), www.rlg.org/preserv/diginews/diginews3-5.html. For an alternative assessment of imaging costs, see the HEDS matrix of potential cost factors in Simon Tanner and Joanne Lomax Smith, "Digitisation: How Much Does It Really Cost?" (paper for the Digital Resources for the Humanities 1999 conference, Kings College London, September 12–15, 1999), heds.herts.ac.uk/HEDCinfo/Papers/drh99.pdf, p. 7. See also Lee, "Scoping the Future," appendix E.

25. John Stokes, "Imaging Pictorial Collections at the Library of Congress," *RLG DigiNews* 3, no. 2 (April 15, 1999), www.rlg.org/preserv/diginews/diginews3-2.html. In the SagaNet project, saga manuscripts are being captured as 25-MB files at 300 dpi, 24-

bit color using the Jenoptik eyelike Digital Camera System; production averages 300 scans per day, leading to the creation of 600 pages per day. Information supplied by Thorstein Halgrimsson, National and University Library of Iceland, to Anne R. Kenney.

26. Production at the Johnson Art Museum at Cornell averages 70 images per day for a nine-hour shift using one digital camera and two photographers. The Museum of Modern Art reports 20 images per day to scan and edit. The equipment used can account for some of the difference. MOMA, for instance, discovered that for different digital scanning backs scan times ranged from 45 seconds to approximately 14 minutes. The staff also discovered that scanning time may not be a true indicator of productivity: color correction took longer for some digital cameras that scanned quickly but had insufficient color tools. Linda Serenson Colet, Kate Keller, and Erik Landsberg, "Digitizing Photographic Collections: A Case Study at the Museum of Modern Art, NY" (paper presented at the Electronic Imaging and the Visual Arts Conference, Paris, September 2, 1997), p. 9.

27. Puglia, "Costs of Digital Imaging Projects."

28. "RLG Tools for Digital Imaging," *RLG*, www.rlg.org/preserv/RLGtools.html.

was derived by Cornell's Department of Preservation from the production environment of a recent in-house imaging effort. These costs are for image acquisition only and do not reflect the costs of providing access to the digitized materials or long-term maintenance.

Cornell identified six cost categories:

- Personnel

- Equipment

- Cataloging

- Supplies

- Contingency

- Overhead/Indirects

Personnel

Personnel includes staff performing specific tasks and a management surcharge for each staff member. Staff costs are calculated using a "weighted hourly rate" which represents over twice the hourly wage:

- 222 work days/year (excluding vacation, sick, personal time, and holidays)

- 7.3 hour work day (excluding breaks)

- 1,621 hours/year at work

- 75% "production time": 1,216 hours/year

- Weighted hourly rate = $\dfrac{\text{salary} + \text{fringe}}{1,216}$

Management costs are a percentage of the supervisor's time "assigned" to each staff member, typically in the 10–25% range at Cornell, depending on the complexity of tasks and staff members' experience.

Equipment

The annual equipment cost is based on:

- The purchase price of hardware and software amortized over the anticipated life span, calculated at 3–5 years.

- Annual maintenance/licenses fees, calculated at 20–50% of equipment purchase price, include onsite maintenance.

- Equipment replacement costs constitute a surcharge of up to 50% of annual equipment cost/year.

- The equipment cost assigned to each digitized item is calculated by dividing the annual equipment, maintenance, and replacement costs by the estimated number of items to be digitized in a year.

Cataloging

Cataloging is the creation of an online catalog record for the digitized item, including maintenance of persistent naming scheme (e.g., PURL).

Supplies and Materials

Supplies and materials include storage media, computer and peripheral supplies, phone/data lines, reference materials, and normal office supplies, increasingly calculated as a 10% levy on other costs.

Contingency

Contingency covers unanticipated expenses incurred in ramp-up, trouble-shooting, training, systems support, etc. This is not traditionally an eligible expense for US funding agencies, but it is increasingly recognized by UK and European funders. Contingency will vary from project to project, depending on complexity, experience of staff, and size of effort.

Overhead/Indirects

Overhead and indirect costs include space, utilities, services, and general and administrative support, calculated on the total direct costs. Cornell's federally negotiated indirect rate for 1999–2003 is 57%, but library-conversion projects are typically charged at 30%. Overhead may be shared by an institution, but it is nonetheless a real expense.

Applying the Cost Model

To illustrate how this approach works, consider a proposed program to digitize brittle books at Cornell. Based on experience and/or time trials, we have estimated production times for a 300-page, text-based book at .75 hours for preparation, .5 hours for quality control, and 1.5 hours for scanning. Preservation assistants perform all preparation and inspection functions (annual salary $18,825 plus benefits calculated at 37%, for a weighted hourly rate of $21.21). A scanning technician is responsible for scanning, rescanning, file naming, document structuring, and related tasks (annual salary $21,375, weighted hourly rate $24.08). A preservation librarian supervises these two staff members and 25% of her time is assigned to managing this effort (annual salary $35,620, weighted hourly rate $40.13). In addition, equipment costs include the production scanner ($17,000) and two workstations with high-end monitors ($3,500 each). Total: $24,000

Personnel (per book)

- Preservation assistant @1.25 hours: $26.51
- Scanning technician @ 1.5 hours: $36.12
- Supervisor (25% of 1.5 hours): $15.05
- TOTAL PERSONNEL: $77.68

Equipment costs (per book)

- At 1.5 hours/book, 811 books can be scanned in 1,216 hours
- Annual scanner and workstation costs: $24,000 amortized over 3 years, or $8,000
- Maintenance/licensing at 30% of purchase price/year: $7,200
- Replacement @ 50% annual equipment cost: $4,000
- Total annual equipment cost: $19,200
- Divided by books per year (811)
- TOTAL EQUIPMENT COST: $23.67

Supplies and Materials (per book)

- 10% of personnel and equipment: $10.14

Cataloging

- Update to NOTIS record to reflect electronic version, including assignment of persistent name (PURL): $10/book

Contingency (per book)

- 10% of other costs ($121.49): $12.15

Overhead/Indirects (per book)

- Total direct costs (other costs + contingency): $133.64
- OVERHEAD/INDIRECT COSTS (30%): $40.09

Total

- Total image-acquisition costs (per book): $173.73
- PER PAGE COST: $0.58

Some Production Figures Compared

In the spring of 1999, the University of Michigan was asked to analyze the range of current cost estimates for digitization, OCR, and encoding of books to assist funding agencies in projecting figures for *large-scale* imaging efforts. Cornell and the University of Virginia evaluated Michigan's figures and, despite some institutional differences, found remarkable agreement on the order of magnitude of basic costs. Table 9.1 shows costs provided by these three institutions based on recent production efforts for scanning disbound books bitonally at 600 dpi, OCRing, and encoding (with and without proofing and correction), with much of the work outsourced. Cost ranges represent institutional differences. *These figures do not include costs of management, cataloging, or institutional overhead.*

Costs of Access and Preservation

The ongoing costs of maintaining and providing access to digital collections outweigh image-acquisition costs. Ongoing costs prove difficult to calculate because the available figures tend to be speculative, reflect start-up costs, and/or focus on the requisite technical infrastructure. A 1991 study conducted by the Environmental Protection Agency pegged the total costs of supplies, services, and hardware to maintain access to digital material for ten years at four to seven times the cost of creation.[29] In 1996, Charles Lowry and Denise Troll estimated that digital files would be 16 times more expensive to maintain and access than their paper counterparts.[30] And at the 1999

29. "Summary for Benefit-Cost Analysis, EPA Superfund Document Management System Concept," 1991.

30. Charles Lowry and Denise Troll, "Virtual Library Project," NASIG Proceedings: Tradition, Technology, and Transformation, Part 1, *Serials Librarian* 28, nos. 1–2 (1996).

NARA/Department of Defense Scanned Images Standards Conference, some participants argued that digital images need to be migrated every three to five years at 50–100% of the costs for the original imaging project.[31] These projections are similar to those reported by the Early Canadiana Online project, which estimated digital storage and access costs at nearly $0.12 per image per year, or 15% of the cost to produce the digital file from microfiche.[32] In the ILEJ project, access and archiving costs were calculated at a little over $0.08/image per year, or nearly 25% of the cost of image creation.[33] Some speculate that costs that would normally decline with greater experience or centralization may actually increase as the supporting technology changes periodically, bringing new financial burdens.[34]

Beyond those costs, we know that digitization leads to increased use of materials, placing new demands on institutional resources of all kinds. Libraries and archives experience incredible responses to digital resources that dwarf use of their physical counterparts. In 1999, the New York Public Library reported ten million hits a month as opposed to 50,000 books served at 42nd Street, and the Library of Congress transmitted over 563 million files.[35] Simply accommodating so many users requires institutions to support extremely powerful access systems as discussed in chapter 7. NARA, for instance, spent nearly as much to upgrade their file servers to provide Web access to the EAP images as they did to

Table 9.1. Digital Conversion Costs (Scanning, OCR, Encoding)

Task	Page Cost	300-Page Book
Preparation	$0.053	$15.90
Scanning and structuring	$0.10–$0.14	$30.00–$42.00
Quality control	$0.035	$10.60
Unproofed OCR (3,000 characters/page)	$0.10	$3.00
OCR proofing and correction	$0.25–$0.60	$75.00–$180.00
Encoding (3,000 characters/page)	$0.40–$0.50	$120.00–$150.00
Proofing of encoded text	$0.44	$132.00
Initial online storage	$0.017	$5.10
Total	$1.40–$1.89	$391.60–$538.60

digitize them, and has delayed further plans for digitization, in part because of the attendant costs.[36]

A growing—and demanding—secondary clientele can tax staff resources as well. The City Archive of Antwerp found that the percentage of non-Belgian users rose from 33% onsite to 55% via their Web site.[37] Cornell's Making of America Web site, consisting of 19th-century journals and monographs, receives 4,000 requests per day. Non-Cornellians make up a very large share of the users and they expect the digital library to act just like a regular library, replete with basic services. As the system has become more stable, user requests have had less to do with system difficulties and more to do with content inquiries, often representing the interests of a general rather than a scholarly audience. Cornell began its digital library a decade ago under the motto "Any time, any place." Today it must address the question, Anybody?

31. Sue MacTavish, "DoD-NARA Scanned Images Standards Conference," *RLG DigiNews* 3, no. 2 (April 15, 1999), www.rlg.org/preserv/diginews/diginews3-2.html. Puglia, "Costs of Digital Imaging Projects," also suggests planning for 50–100% of initial cost per image for maintenance for the first ten years. Also see Anne R. Kenney, *Digital to Microfilm Conversion: A Demonstration Project, 1994–1996, Final Report* (Ithaca, NY: Cornell University Library, 1997); available from "Publications," *Cornell University Library, Preservation & Conservation,* www.library.cornell.edu/preservation/pub.htm.

32. Bruce Kingma, *The Economics of Digital Access: The Early Canadiana Online Project,* eco480.bdraper.albany.edu/ECO/kingma_report.PDF.

33. "ILEJ Final Report." (Exchange rate £1 = $1.6526 US).

34. Howard Besser, "Digital Image Distribution: A Study of Costs and Uses," *D-Lib Magazine* 5, no. 10 (October 1999), www.dlib.org/dlib/october99/10besser.html.

35. Robert Darnton, "The New Age of the Book," *The New York Review of Books* (March 18, 1999): pp. 5–7; "Library of Congress WWW Usage Statistics," *Library of Congress,* lcweb.loc.gov/stats.

36. Puglia, "Costs of Digital Imaging Projects"; "Testimony Submitted by H. Thomas Hickerson to the Committee on Government Reform's Subcommittee on Government Management, Information, and Technology: Oversight Hearing on the National Archives and Records Administration, October 20, 1999," *The Society of American Archivists,* www.archivists.org/statements/hickerson_testimony.html.

37. "Proposal," *European Visual Archive,* www.eva-eu.org/proposal.htm.

IS DIGITIZATION ECONOMICALLY VIABLE?

Most digital conversion projects have been funded by one-time appropriations from government, foundation, or institutional sources. Ultimately, an institution must face the ongoing costs of maintaining its digital assets, which will require the commitment of institutional resources. Fund reallocation is not a pleasant prospect for any institution, so there is a great deal of interest in considering the economic sustainability of digital image conversion programs. Indeed, one of the main advantages often cited for digitization in the early 1990s was the potential to curb the spiraling costs of managing and making accessible scholarly information.[38] Institutions can accomplish this two ways: by diverting resources saved in other operational areas to the digital effort, or by passing costs on to users of digital products.

Reducing Institutional Expenses

A number of studies have compared the total costs of creating and maintaining digital collections to the cost of acquiring and maintaining print collections. Many of these studies have focused on current resources, especially scholarly journals, and a number of them have factored in editorial and publication costs as well.[39] Several libraries are attempting to be "electronic only," including the Technical Knowledge Center and Library of Denmark. The Center credits the savings from no longer handling the print journals for increasing the number of titles it provides—estimated at 25% more than was possible in the print realm.[40] It is still too early to tell whether such electronic-only libraries will meet all the information needs of their users, result in significant cost savings over time, and be able to ensure continuing long-term access to electronic information.

Government agencies such as the Army Corps of Engineers, the Naval Research Lab, and the US Patent Office can demonstrate overall cost savings by converting their entire back files to digital form and installing automated document-handling systems. The Naval Research Lab, which converted its collection of unclassified documents representing nearly nine million pages, calculated that the disposition of the originals has resulted in space savings of over $100,000/year. The Army Corps study cites primarily the more rapid location and delivery of documents within the agency and to regional offices and the reduced need for copying and duplicating records.[41]

Unfortunately, many libraries and archives do not share the circumstances for these cost savings:

▸ Government agencies own the material in their collection without copyright constraint, whereas a great percentage of library and archive holdings is covered by copyright, privacy, or donor restriction.

▸ Government agencies are top-down, hierarchical operations, where decisions associated with collection retention or technological change involve the whole agency. Cultural repositories may not be in a similar position. The library system at most universities, for example, is in no position to impose such changes on the faculty.

▸ Government collections are unique and self-contained, with little concern for external constituencies; most libraries and archives serve large, heterogeneous user populations.

▸ The primary users of the documents in government agencies are employees and agency staff

38. See in particular Anthony M. Cummings et al., *University Libraries and Scholarly Communication: Study Prepared for The Andrew W. Mellon Foundation* (Washington, DC: Association of Research Libraries, 1992).

39. Andrew Odlyzko, "Competition and Cooperation: Libraries and Publishers in the Transition to Electronic Scholarly Journals" (revised version April 17, 1999), www.research.att.com/~amo/doc/competition.cooperation.pdf, and Odlyzko, "Economics of Electronic Publishing"; Richard De Gennaro, "JSTOR: Building an Internet Accessible Digital Archive of Retrospective Journals" (63rd IFLA General Conference, Conference Programme and Proceedings, August 31–September 5, 1997), *IFLANET*, ifla.inist.fr/IV/ifla63/63genr.htm. Michael Lesk, *Practical Digital Libraries: Books, Bytes, and Bucks* (San Francisco: Morgan Kaufmann Publishers, Inc., 1997).

40. Katie Hafner, "Scientists are Publishing More On Line," *The New York Times* (January 21, 1999): p. D4. "Technical Knowledge Center & Library of Denmark Revolutionises the Library World," *Technical Knowledge Center & Library of Denmark*, www.dtv.dk/new/press/210199_e.htm. For other electronic-only efforts, see "Fiterman Hall Virtual Library & Study Rooms," *Borough of Manhattan Community College*, www.bmcc.cuny.edu/lib/ virtue.html, and "Mission Statement," *Cal State Monterey Bay, Library*, library.monterey.edu/about/sub_mission.html.

41. Patricia A. Ames, acting deputy librarian, Naval Research Lab, e-mail to Richard Entlich, September 3, 1999. US Army Corps of Engineers, "Economic Analysis, Executive Summary Report, Optical Disk Imaging," July 25, 1996.

do most document retrieval. While the same may apply to many archives, in most libraries users will benefit more than staff from faster document retrieval.

The potential for cost savings is at the heart of one of the largest digitization projects today. JSTOR, a nonprofit organization digitizing back issues of scholarly journals, takes as its premise that space savings is a key factor for libraries. JSTOR contends that a single library cannot save money by digitizing its older holdings, but cooperative, multilibrary agreements might be economical. In addition, JSTOR hopes to provide a trusted source to maintain the archive of important scholarly journal literature going forward.

JSTOR's cost assessment assumes libraries can pay for their subscription fees by discarding paper holdings or moving them to cheaper, less accessible offsite storage. Libraries presumably would free up space for other materials, reduce the need to build new libraries, and incur additional operational savings in binding, preservation, retrieval, reshelving, staff redeployment, etc. This model assumes that readers will prefer to use journals in their electronic form and that libraries can trust JSTOR to maintain its digital holdings in perpetuity. Use figures certainly attest to reader interest in electronic access: in 1999, JSTOR reported over 16 million "significant accesses."[42] But to date, the promise of cost savings has yet to be realized, as few institutions if any have taken steps to wean themselves from the hard-copy versions or even to move physical volumes to offsite storage. For the time being, it seems that JSTOR members are subscribing for the enhancements and convenience and thus have increased rather than decreased expenses.[43]

Of course, JSTOR has its own expenses. The Mellon Foundation made grants exceeding $4 mil-

lion to establish JSTOR and it is estimated that once JSTOR reaches its initial goal of back runs of 100 journals, the annual operating expenses will total $2.5 million (which includes a staff of 17). Presumably, this figure will increase as additional titles are added. Based on the sliding scale of annual access fees to JSTOR subscribers, a total of $4.676 million could be raised annually if all 1,402 US higher-education institutions subscribed (there are currently over 500 subscribers). JSTOR will have to increase its membership base by an additional 50% (the exact proportion would depend on the distribution of large, medium, small, and very small institutions) in order to break even on the annual operating expenses for 100 journals. It is still too early to know whether this extraordinary enterprise in particular will be sustainable when member institutions eventually face the economic difficulties of continuing to support both the traditional and the electronic collections. This model offers compelling—and inevitably triumphant—economic and access advantages, but it appears that the transition from print to electronic will take longer than originally expected.[44]

Recovering Costs

Although most institutions do not digitize to make money, some hope to cover costs by generating revenue, which conflicts with many users' assumptions that everything on the Web should be free. Indeed, many institutions currently provide free access to at least base-level images, in part because they have received outside funds and in part because their administration has borne the short-term expense of developing an electronic presence. As institutions face the need to fund digital efforts internally, there will be growing pressure to recover costs.[45] Cost-recovery solutions have been

42. "JSTOR Usage Statistics," stats.jstor.org.

43. "The Need," *JSTOR*, www.jstor.org/about/need.html; Kevin M. Guthrie, "JSTOR: The Development of a Cost-Driven, Value-Based Pricing Model" (paper presented at ARL's Scholarly Communication and Technology conference, Emory University, April 24–24, 1997), *Association of Research Libraries*, arl.cni.org/scomm/scat/guthries.html; William G. Bowen "JSTOR and the Economics of Scholarly Communication" (October 4, 1995), *The Andrew W. Mellon Foundation, Journal Storage Project*, www.mellon.org/jsesc.html; De Gennaro, "JSTOR: Internet Accessible Digital Archive." JSTOR's late-1999 survey should include feedback on cost recovery through JSTOR participation. In the NEH-funded preservation microfilming projects, 65–95% of the microfilmed brittle books go back on the shelves, either in the main

library or off-site; Jeff Field, "NEH Project Managers Discuss the Hybrid Approach for Digitization and Preservation Microfilming" (National Endowment for the Humanities, Division of Preservation and Access, October 25, 1999).

44. DeGennaro, "JSTOR: Internet Accessible Digital Archive," and Guthrie, "JSTOR: Pricing Model."

45. Seven percent of libraries and archives responding to an international survey cited commercial exploitation as a selection criteria for digitization; Gould and Ebdon, "IFLA/UNESCO Survey," p. 13. The economics of managing intellectual property in a networked environment are discussed in Diane M. Zorich, *Introduction to Managing Digital Assets: Options for Cultural and Educational Organizations* (Los Angeles: Getty Information Institute, 1999).

advanced, but to date there is little hard evidence that they will succeed.

▸ In 1997, The British Library developed a business case to seek private sector collaboration in creating a self-sustaining digital library service. Unfortunately, after a year of negotiation, the library and the bidding consortium agreed to discontinue negotiations, as it proved impossible to "balance the objectives of the Library with the commercial operating requirements of the consortium."[46]

▸ The Scottish Millennium Commission (which received funds from the Scottish National Lottery) funded the Scottish Cultural Resource Access Network (SCRAN) to develop a self-financing organization by 2001. SCRAN's initial efforts have focused on the development of a sizeable database of text records of historic monuments and artifacts held in museums, galleries, and archives. Copyright licenses for member and nonmember access will provide a critical source of anticipated income.[47]

▸ The Early Canadiana Online project conducted a feasibility study on the economics of production, storage, and distribution of digitized versions for over 3,000 titles from the Canadian Institute of Historical Microreproductions collection. It showed that the costs of digital information would be lower than microfiche or hard copies for both libraries and patrons—given a sufficient subscription base. Nonetheless, the report concluded that the costs of creation, maintenance, and access would require a combination of "value based pricing of information as well as the solicitation of grants and donations."[48]

▸ The Pricing Electronic Access to Knowledge (PEAK) project, conducted at the University of Michigan over an 18-month period ending in August 1999, tested a range of access models for delivering 1,200 Elsevier Science electronic journals to a dozen institutions, including various subscription-based models. Although the

project was fairly short-term and the findings inconclusive, it did not break even. Expenditures were estimated at $400,000; participating institutions' fees covered 45% of the costs, with the remaining expenses covered by in-kind contributions from Elsevier and the university.[49]

▸ The Digital Scriptorium (DS) is a joint project of UC Berkeley and Columbia University to enhance scholarly access and make illuminated manuscripts more widely available for education purposes. The project assumes, however, that researchers must bear some of the costs and that they may well support this if it can be demonstrated that the Digital Scriptorium will save them time and money. DS is examining the potential user audience, how much various segments of that audience might be willing to pay for access, how to respond to specific research needs, and ways to forecast future use and revenues.[50]

▸ In 1997, the MAGNETS project conducted by VASARI issued a report on general market and economic aspects of electronic access to European museum holdings, urging greater collaboration between museums and the business sector. Calling for further study, the report concluded, however, that sustainable electronic commerce awaits the resolution of technical issues associated with rights management, pricing, billing, and payment. One of the more interesting findings was that "currently there are grounds to believe that after initially undervaluing museums' electronic rights, these are now being overvalued by some museums and that the pendulum will swing back to a more moderate position—this may take 3–5 years."[51]

▸ The Museum Educational Site Licensing Project (MESL) attempted to address many of the issues

46. The British Library home page, www.bl.uk, link to Digital Library.

47. "Information," SCRAN home page, www.scran.ac.uk/faq/quest1.htm.

48. Kingma, *Economics of Digital Access.*

49. Jeffrey K. MacKie-Mason et al., "A Report on the PEAK Experiment: Usage and Economic Behavior," *D-Lib Magazine* 5, no. 7/8 (July/August 1999), www.dlib.org/dlib/july99/mackie-mason/07mackie-mason.html.

50. Malcolm Getz, "The Digital Scriptorium: Medieval Manuscripts on the World Wide Web" (March 16, 1998), www.vanderbilt.edu/Econ/faculty/Getz/DSvision.pdf, and Getz, questionnaire for Digital Scriptorium users, www.vanderbilt.edu/Econ/faculty/ Getz/DSquery.pdf.

51. James Helmsley, ed., "MAGNETS Museum and Galleries New Technology Study: Market and General Economic Issues," (January 1997) *VASARI,* www.vasari.co.uk/magnets/wp4.

Sustainability through Integration
Susan M. Yoder
Director, Integrated Information Services, Research Libraries Group

Most institutions face the desire to digitize with a modicum of money. In many instances, the funding for these projects will not cover the entire cost of digitizing the initial corpus, much less keep an ongoing service in the black. Additionally, some institutions are interested in generating a new revenue stream by providing enhanced access to their historical assets. This is made more difficult by the widely held perception that much of this information is, and should be, freely available. These tensions mix the cultural and commercial worlds more closely than ever. So, what issues should you review in assessing the viability of digital imaging initiatives? A business analysis at the front end will lessen the chance that your work will end up in the "dead projects" room at the back end.

Know Your Purpose
Why are you embarking on this endeavor? Is it for preservation, or use, or generation of income, or some combination of reasons? An imaging project may not serve all purposes. If, for instance, you want to emphasize broad access, and others in your institution want a significant revenue stream, you have positioned yourself directly in the cross hairs. It is not enough to get consensus that digitizing is good, or that you should do it, or even which collections to target. Everyone involved should agree (or at least be aware of) why it is being done.

Do Not Digitize Solely to Generate Revenue
The best way to ensure success is to identify multiple benefits. In the current environment, most institutions and organizations find it difficult to create a wholly self-sustaining image service that not only eliminates the need for continual outside funding, but also generates incremental dollars for the communal till. Unless you are willing to step totally into the commercial world, you should have at least one compelling reason for digitizing your collection other than generating revenue.

Know Your Market
If you are digitizing a specific collection at an academic institution for your own in-house classroom use, you are no doubt well aware of how, when, and why your clientele will use these resources. If, on the other hand, you are at a historical society, an archive, a museum, or a university with multiple, disparate collections, it is quite possible that you could not easily define every possible user and use of your materials. Without some up-front analysis of your users' needs—and potential groups to reach—not only can you not set your prices, but you also cannot adequately design your interface. Nor can you make decisions about content, determine how to communicate information most effectively, or set up the best delivery and payment mechanisms.

Do not assume that your collections and your chosen access methodology will be appropriate for every possible type of user. Many organizations have weakened the potential of their offerings by attempting to build a universal service, hoping to be "all things to all people."

Understand Your Expertise
It is seductive to think that the images in and of themselves are your highly valuable commodity. Increasingly, however, easily navigable sites with supportive information are what make images useful. As you decide whether, what, and how much to digitize, keep in mind that your information *about* those images, when integrated with them, can often decide someone in favor of paying for access to your collections.

Establish Your Control Comfort Zone
As authentication improves, controlling specific uses of images will be easier. You need to determine early in the digitization process exactly how far your rights go in allowing you to disseminate your images, and weigh your interest in revenue against what uses (and users) your organization feels comfortable in supporting. For example, you may decide to allow unrestricted access to onscreen views, but control printing. This dual approach may mollify those who expect all networked information to be free. In addition, you may decide that increased revenue from nonacademic users is well worth the additional authentication, but find that you need to develop authorization mechanisms in order to control commercial use of certain images.

Test Your Price Assumptions
If you think there is a viable market for your image resources, you still need to determine their fair market value and the best means for securing revenue. Image-based services are generally priced by subscription, pay per view, or some combination of the two. For established text databases, a subscription that covers the entire service is by far the most tried-and-true model and one with which most academic institutions are comfortable for many reasons, not the least being budget control. However, image services are creating different use patterns, and until these new behaviors are better understood, setting an up-front, unlimited-use subscription price can be difficult.

Many commercial image distributors and museums favor the pay-per-view approach. Generally, searching is free and a charge is incurred only for a specific transaction, such as asking to see a higher-resolution image, ordering a print, or requesting approval for a specific use of a particular image. Unless, however, you have a separate, established revenue stream to underwrite the first year of your service, basing initial revenue estimates on per-view transactions is risky. It depends on knowledge of end-user behavior that usually is just not available at the beginning of these initiatives.

Fair market value is usually determined by analyzing similarly positioned services, plus determining the price flexibility of your

CONTINUED ON PAGE 174

CONTINUED FROM PAGE 173

user base. The newer the market, the more difficult these two analyses are; however, you should look at services that are as comparable as possible in both type of image and type of user. Although questionnaires can help determine how flexible users are about actual prices, it is much more accurate to analyze increases or decreases in use levels for a particular service following a price change.

Identify Potential Partners

One of the best ways to ensure a return on your investment is to join with other institutions. Collective projects have a number of benefits:

▸ *Achieving critical mass*: A financially viable service is one its users consider necessary. For this, the site needs to always have available the images that the user seeks. Working with other institutions that have complementary collections greatly increases the chances of providing successful search results. A larger corpus of material also broadens the potential use of a specific collection. A service or site can be the basic component of an introductory art class, for example, only if the content is appropriately inclusive.

▸ *Leveraging additional data and services*: Linking with other data and media can be crucial, leading to an extensive network not only of images but also of library catalogs, union catalogs, slide libraries, citation databases, full text, and other reference resources. Cooperation tends to lead to wider dissemination, better systems, and earlier payback.

▸ *Negotiating volume discounts*: For outsourcing some of the basic production work, a group initiative would more likely be eligible for volume discounts.

The RLG/AMICO Collaboration

In July 1999, RLG and the AMICO Consortium began to offer access on the Web to the AMICO Library™, a digital library documenting museum collections from multiple institutions, and RLG is expanding the site with additional individual and joint collections of cultural heritage materials. The partnership with AMICO was logical: RLG offers stable, respected research services worldwide, and the AMICO Consortium (headed, with great organizational skill, by Jennifer Trant and David Bearman) has constructed not only a focused, high-quality database, but a working methodology for rights management. The success of this project demonstrates the value of paying attention to all aspects of a service, from initial image collection through to final delivery. And most importantly, it has shown the strength of working collaboratively—integrating images *and* institutions—to create a financially viable service.

Note

1. AMICO: Art Museum Image Consortium home page, www.amico.net.

related to consortial licensing of museum images to universities for educational purposes. A detailed financial assessment concluded that consortial distribution of digitized museum objects to educational institutions will likely not be economically self-sustaining or revenue producing for some years to come.[52]

▸ AMICO (Art Museum Image Consortium) built on the MESL experience and in collaboration with RLG is moving the concept of consortial licensing of museum images to educational institutions one step closer to large-scale reality. The project focuses on taking advantage of emerging educational opportunities, but there is also a clear expectation that it will bring new revenue sources and greater economic stability to the participating museums.[53] It is unclear at this point what impact the Academic Image Exchange, sponsored by the Digital Library Federation and the College Art Association to provide free access to art images for educational and nonprofit use, will have on AMICO's market.[54]

TOWARDS A NEW PARADIGM

Given the current environment, digitization will not pay for itself—at least not in the short run—and cultural institutions should not embark on the transition from individual projects to ongoing programs unless they are persuaded that digitization

51. James Helmsley, ed., "MAGNETS Museum and Galleries New Technology Study: Market and General Economic Issues" (January 1997), *VASARI*, www.vasari.co.uk/magnets/wp4.

52. Howard Besser and Robert Yamashita, "The Cost of Digital Image Distribution: The Social and Economic Implications of the Production, Distribution, and Usage of Image Data, Final Report," (1998) sunsite.Berkeley.edu/Imaging/Databases/1998mellon. One collaborative initiative stemming from the MESL experience, The Museum Digital Library Collection, Inc., aspired to become a nationwide image-licensing enterprise, but has become moribund; Museum Digital Library Collection home page, www.museumlicensing.org.

53. David Bearman, "New Economic Models for Administering Cultural Intellectual Property" (1996), *Archives and Museum Informatics*, www.archimuse.com/papers/db.mesl/economics.html. Jennifer Trant and David Bearman, "The Art Museum Image Consortium: Licensing Museum Digital Documentation for Educational Use," *Archives & Museum Informatics* (fall 1997), www.archimuse.com/papers/amico.spectra.9708.html.

54. Rebecca Graham, "Academic Image Exchange Formed," *CLIR Issues* 10 (July/August 1999), www.clir.org/pubs/issues/issues10.html. In September 1999, The Andrew W. Mellon Foundation awarded a grant to the Council on Library and Information Resources to develop the project.

serves their essential mission in very compelling ways. If they are not, institutions must still address the information revolution, the changing expectations of their clientele, and the increasing costs of managing underutilized historical materials. If institutions are convinced of the value of digitization, their efforts may have a greater chance of becoming sustainable, *if*:

▸ Institutions treat digital material as critical assets and invest wisely in the selection and creation of digital resources that are likely to be used and reused over time.

▸ Digital projects become digital programs supported at the highest levels and provided a secure institutional base.

▸ Committed resources fund digital programs. Because digital assets must be assured of perpetual care, a social security fund should be established from institutional resources.

▸ Digital programs encompass the full life cycle of digital objects, including provisions for their long-term management. Preservation concerns should be addressed from the ground up, following recommendations presented in chapter 8. Unless preservation is considered at the point of creation, "there is little prospect of archiving image resources that will survive technological change."[55]

55. Michael Ester, *Digital Image Collections: Issues and Practice* (Washington, DC: Commission on Preservation and Access, 1996), p. 16.

▸ Online access systems do not jeopardize digital assets by applying short-term solutions to short-term problems. As John Price-Wilkin noted in chapter 6, many of today's constraints will not be tomorrow's and we should not build an approach that becomes quickly outdated or superceded.

▸ Institutions substitute digital resources for some traditional resources and services, and/or researchers embrace the use of digital image collections and are willing to pay part of the cost for the added value or convenience that they offer.

▸ Institutions are prepared to cooperate in sharing the rewards and responsibilities of the digital world. Meaningful transition can not happen on just the local level.

Libraries and archives should view digital conversion as a means to other goals, not an end in itself. Comparing digitization to traditional resources and services may not be the best way to judge its value. Through digitization, an institution might enable new forms of scholarship or teaching, breathe new life into older materials, introduce efficiencies and streamline services, or protect institutional assets while increasing access. Under these and similar circumstances, digitization may be a legitimate loss leader that results in a new service paradigm, enabling libraries and archives to meet the changing needs of their primary clientele and compete successfully with other information providers in reaching a broad range of cultural consumers.

Index